Human Performance Measures Handbook

Human Performance Measures Handbook

Valerie J. Gawron
Veridian Engineering

LEA LAWRENCE ERLBAUM ASSOCIATES, PUBLISHERS
2000 Mahwah, New Jersey London

The material on Measures for Human Performance is based on Appendix B of ANSI/AIAA G—035-1992, *Guide for Human Performance Measurements*, American Institute of Aeronautics and Astronautics, Reston, VA, copyright, 1993.

The final camera copy for this work was prepared by the author, and therefore the publisher takes no responsibility for consistency or correctness of typographical style. However, this arrangement helps to make publication of this kind of scholarship possible.

Lawrence Erlbaum Associates, Inc., Publishers
10 Industrial Avenue
Mahwah, New Jersey 07430-2262

Cover design by Kathryn Houghtaling Lacey

Library of Congress Cataloging-in-Publication Data

Gawron, Valerie, J.
Human performance measures handbook / Valerie J. Gawron.
p. cm.
Includes bibliographical references and indexes.
ISBN 0-8058-3701-9 (pbk. : alk. paper)
1. Human engineering—Handbooks, manuals, etc. 2. Human–machine systems—Handbooks, manuals, etc. I. Title.

T59.7.G38 2000
620.8'2—dc21 00-028774

Books published by Lawrence Erlbaum Associates are printed on acid-free paper, and their bindings are chosen for strength and durability

Printed in the United States of America
10 9 8 7 6 5 4 3 2 1

Dedication

To my Dad, Stanley C. Gawron, 17 March 1921 to 9 February 2000.

Contents

List of Figures

List of Tables

Preface

This human performance measures handbook was developed to help researchers and practitioners select measures to be used in the evaluation of human/machine systems. It can also be used to supplement classes at both the undergraduate and graduate courses in ergonomics, experimental psychology, human factors, human performance, measurement, and system test and evaluation. The handbook begins with an overview of the steps involved in developing a test to measure human performance, workload, and/or situational awareness. This is followed by a definition of human performance and a review of human performance measures. Workload and Situational Awareness are similarly treated in subsequent chapters.

Acknowledgments

This book began while I was supporting numerous test and evaluation projects of military and commercial transportation systems. Working with engineers, operators, managers, programmers, and scientists showed a need for both educating them on human performance measurement and providing guidance for selecting the best measures for the test. I thank my team members for their patience and openness. I also thank Dr. Dave Meister who provided great encouragement to me to write this book based on his reading of my "measure of the month" article in the Test and Evaluation Technical Group newsletter. He and Dr. Tom Enderwick also provided a thorough review of the first draft of this book. For these reviews I am truly grateful.

1 Introduction

Human factors specialists, including industrial engineers, engineering psychologists, human factors engineers, and many others, consummately seek better (more efficient and effective) ways to characterize and measure the human element as part of the system so we can build trains, planes, and automobiles with superior human–system interfaces. Yet the human factors specialist is often frustrated by the lack of readily accessible information on human performance, workload, and Situational Awareness (SA) measures. This book guides the reader through the critical process of selecting the appropriate measures of human performance, workload, and SA and later, provides specific examples of such.

There are two types of evaluations of human performance. The first type is subjective methods. These are characterized by humans providing opinions through interviews and questionnaires or by observing others' behavior. There are several excellent references on these techniques (e.g., Meister, 1986). The second type of evaluation of human performance is the experimental method. Again there are several excellent references (e.g., Keppel, 1991; Kirk, 1995). This experimental method is the focus of this book. Chapter 1 is a short tutorial on the experimental design; Chapter 2 describes measures of human performance; Chapter 3, measures of workload; and Chapter 4, measures of SA.

For the tutorial, the task of selecting between aircraft cockpit displays is used as an example. For readers familiar with the general principles of experimentation, this should be simply an interesting application of academic theory. For readers who may not be so familiar, it should provide a good foundation of why it is so important to select the right measures in preparation of carrying out your experiment.

The need for efficient and effective selection of the appropriate human performance, workload, and SA measures has never been greater. However, little guidance has been provided to support this selection process. This book was written to meet this need. The book begins with an example in which an experimenter must select measures of performance and workload to evaluate a cockpit display. Next, human performance is defined and measures presented. Each measure is described, along with its strengths and limitations, data requirements, threshold values, and sources of further information. After all the performance measures are described, a procedure for selecting among them is presented. In the last section, workload is defined and workload measures described in the same format as performance measures. To make this desk reference easier to use, extensive author and subjective indices are provided.

1.1 The Example

An experiment is a comparison of two or more ways of doing things. The "things" being done are called *independent variables*. The "ways" of doing things are called *experimental conditions*. The measures used for comparison are *dependent variables*. Designing an experiment requires: defining the independent variables, developing the experimental conditions, and selecting the dependent variables. Ways of meeting these requirements are described in the following steps.

Step 1: Define the Question

Clearly define the question to be answered by the results of the experiment. Let's work through an example. Suppose a moving map display is being designed and the lead engineer wants to know if the map should be designed as track up, north up, or something else. He comes to you for an answer. You have an opinion but no hard evidence. You decide to run an experiment. Start by working with the lead engineer to define the question. First, what are the ways of displaying navigation information, that is, what are the experimental conditions to be compared? The lead engineer responds, "Track up, north up, and maybe something else". If he can not define something else, you can not test it. So now you have two experimental conditions: track up versus north up. These conditions form the two levels of your first independent variable, direction of map movement.

Step 2: Check for Qualifiers

Qualifiers are independent variables that qualify or restrict the generalizability of your results. In our example, an important qualifier is the type of user of the moving map display. Will the user be a pilot (who is used to track up) or a navigator (who has been trained with north-up displays)? If you run the experiment with pilots, the most you can say from your results is that one type of display is best *for pilots*. There is your qualifier. If your lead engineer is designing moving map displays for both pilots and navigators, you have only given him half an answer; or worse, if you did not think about the qualifier of type of user, you may have given him an incorrect answer. So check for qualifiers and use the ones that will have an effect on decision making, as independent variables.

In our example, the type of user will have an effect on decision making, so it should be the second independent variable in the experiment. Also in our example, the size of the display will not have an effect on decision making since the lead engineer only has room for an 8-inch display in the instrument panel. Therefore, size of the display should not be included as an independent variable.

Step 3: Specify Conditions

Specify the exact conditions to be compared. In our example, the lead engineer is interested in track up versus north up. So the movement of the map will vary between the two conditions, but everything else about the displays (e.g., scale factor, display resolution, color quality, size of the display, and so forth) should be exactly the same. This way, if the subjects' performance using the two types of displays is different, that difference can be attributed only to the type of display and not to some other difference between the displays.

Step 4: Match Subjects

Match the subjects to the end users. If you want to generalize the results of your experiment to what will happen in the real world, try to match the subjects to the users of the system in the real world. This is extremely important since subjects' past experiences may greatly affect their performance in an experiment. In our example, we added a second independent variable to our experiment specifically because of subjects' previous experiences (that is, pilots are used to track up, navigators are trained with north up). If the end users of the display are pilots, we should use pilots as our subjects. If the end users are navigators, we should use navigators as our subjects. Other subject variables may also be important; in our example, age and training are both very important. Therefore, you should identify what training the user of the map display must have and provide that same training to the subjects before the start of data collection.

Age is important because pilots in their 40s may have problems focusing on near objects such as map displays. Previous training is also important: F-16 pilots have already used moving map displays while C-130 pilots have not. If the end users are pilots in their 20s with F-16 experience, and your subjects are pilots in their forties with C-130 experience, you may be giving the lead engineer the wrong answer to his question of which type of display is better.

Step 5: Select Performance Measures

Your results are influenced to a large degree by the performance measures you select. Performance measures should be relevant, reliable, valid, quantitative, and comprehensive. Let's use these criteria to select performance measures for our example problem.

Criteria 1: Relevant. Relevance to the question being asked is the prime criteria to be used when selecting performance measures. In our example, the lead engineer's question is "What type of display format is better?" Better can refer to staying on course better (accuracy) but it can also refer to getting to the waypoints on time better (time). Subjects' ratings of which display format they prefer does not answer the question of which display is better from a performance standpoint because preference ratings can be affected by factors other than performance.

Criteria 2: Reliable. Reliability refers to the repeatability of the measurements. For recording equipment, reliability is dependent on careful calibration of equipment to ensure that measurements are repeatable and accurate; (i.e., an actual course deviation of 50.31 feet should always be recorded as 50.31 feet). For rating scales, reliability is dependent on the clarity of the wording. Rating scales with ambiguous wording will not give reliable measures of performance. For example, if the question on the rating scale is "Was your performance okay?" the subject may respond "No" after his first simulated flight but "Yes" after his second, simply because he is more comfortable with the task. If you now let him repeat his first flight, he may respond, "Yes." In this case, you are getting a different answer to the same question in the same condition. Subjects will give more reliable responses to less ambiguous questions such as "Did you deviate more than 100 feet from course in this trial?" Even so, you may still get a first "No" and a second "Yes" to the more precise question, indicating that some learning had improved his performance the second time.

Subjects also need to be calibrated. For example, if you are asking which of eight flight control systems is best and your metric is an absolute rating (e.g., Cooper-Harper Handling Qualities Rating), your subject needs to be calibrated with both a "good" aircraft and a "bad" aircraft at the beginning of the experiment. He may also need to be recalibrated during the course of the experiment. The symptoms that suggest the need to recalibrate your subject are the same as those that indicate that you should recalibrate your measuring equipment: (a) all the ratings are falling in a narrower band than you expect, (b) all the ratings are higher or lower than you expect, and (c) the ratings are generally increasing (or decreasing) across the experiment independent of experimental condition. In these cases, give the subject a flight control system that he has already rated. If this second rating is substantially different from the one he previously gave you for the same flight control system, you need to recalibrate your subjects with an aircraft that pulls their ratings away from the average: bad aircraft if all the ratings are near the top, good aircraft if all the ratings are near the bottom.

Criteria 3: Valid. Validity refers to measuring what you really think you are measuring. Validity is closely tied to reliability. If a measure is not reliable, it can never be valid. The converse is not necessarily true. For example, if you ask a subject to rate his workload from 1 to 10 but do not define for him what you mean by workload, he may rate the perceived difficulty of the task rather than the amount of effort he expended in performing the task.

Criteria 4: Quantitative. Quantitative measures are easier to analyze than qualitative measures. They also provide an estimate of the size of the difference between experimental conditions. This is often very useful in performing trade-off analyses of performance versus cost of system designs. This criterion does not preclude the use of qualitative measures, however, because qualitative measures often improve the understanding of experiment results. For qualitative measures, an additional issue must be considered - the type of rating scale. Nominal scales assign an adjective to system being evaluated, (e.g., easy to use). "A nominal scale is categorical in nature, simply identifying differences among things on some characteristic. There is no notion of order, magnitude or size" (Morrow, Jackson, Disch, and Mood, 1995, p. 28). Ordinal scales rank systems being evaluated on a single or a set of dimensions (e.g., the north-up is easier than the track-up display). "Things are ranked in order, but the difference between ranked positions are not comparable" (Morrow, Jackson, Disch, and Mood, 1995, p. 28). Interval scales have equal distances between the values being used to rate the system under evaluation. For example, a bipolar rating scale is used in which the two poles are *extremely easy to use* and *extremely difficult to use*. In between these extremes are the words *moderately easy, equally easy,* and *moderately difficult*. The judgment is that there is an equal distance between any two points on the scale. The perceived difficulty difference between *extremely* and *moderately* is the same as between *moderately* and *no difference*. However, "the zero point is arbitrarily chosen" (Morrow, Jackson, Disch, and Mood, 1995, p. 28). The fourth type of scale is a ratio scale which possesses a true zero (Morrow, Jackson, Disch, and Mood, 1995, p. 29). More detailed descriptions of scales are presented in Baird and Noma (1978), Torgerson (1958), and Young (1984).

Criteria 5: Comprehensive. Comprehensive means the ability to measure all aspects of performance. Recording multiple measures of performance during an experiment is cheaper than setting up a second experiment to measure something that you missed in the first experiment. So measure all aspects of performance that may be influenced by the independent variables. In our example, subjects can trade off accuracy for time (e.g., cut a leg to reach a waypoint on time) and vice versa (e.g., go slower to stay on course better), so we should record both accuracy and time measures.

Step 6: Use Enough Subjects

Use enough subjects to statistically determine if there is a difference in the values of the dependent variables between the experimental conditions. In our example, is the performance of subjects using the track-up display versus the north-up display statistically different? Calculating the number of subjects you need is very simple. First, predict how well subjects will perform in each condition. You can do this using your own judgment, previous data from similar experiments, or from pretest data using your experimental setup. In our example, how much error will there be in waypoint arrival times using the track-up display and the north-up display? From previous studies, you may think that the average error for pilots using the track-up display will be 1.5 seconds and using the north-up display, 2 seconds. Similarly, the navigators will have about 2 seconds error using the track-up display and 1.5 seconds error with the north-up display. For both sets of subjects and both types of displays, you think the standard deviation will be about 0.5 second.

Now we can calculate the effect size, that is, the difference between performances in each condition:

$$\text{effect size} = \frac{|\text{performance in track up - performance in north up}|}{\text{standard deviation}}$$

$$\text{effect size for pilots} = \frac{|1.5 - 2|}{0.5} = 1$$

$$\text{effect size for navigators} = \frac{|2 - 1.5|}{0.5} = 1$$

In Figure 1 we can now read the number of subjects needed to discriminate the two conditions. For an effect size of 1, the number of subjects needed is 18. Therefore we need 18 pilots and 18 navigators in our experiment. Note that although the function presented in Figure 1 is not etched in stone, it is based on over 100 years of experimentation and statistics.

Note that you should estimate your effect size in the same units as you will use in the experiment. Also note that because effect size is calculated as a ratio, you will get the same effect size (and hence the same number of subjects) for equivalent measures. Finally, if you have no idea of the effect size, try the experiment yourself and use your own data to estimate the effect size.

Step 7: Select Data-Collection Equipment

Now that you know the size of the effect of the difference between conditions, check that the data-collection equipment you have selected can reliably measure performance at least one order of magnitude smaller than the smallest discriminating decimal place in the size of the expected difference between conditions. In our example, the expected size in one condition was 1.5 seconds. The smallest discriminating decimal place (1.5 vs. 2.0) is tenths of a second. One order of magnitude smaller is hundredths. Therefore the recording equipment should be accurate to 1/100th of a second. In our example, can the data collection equipment measure time in hundredths of a second?

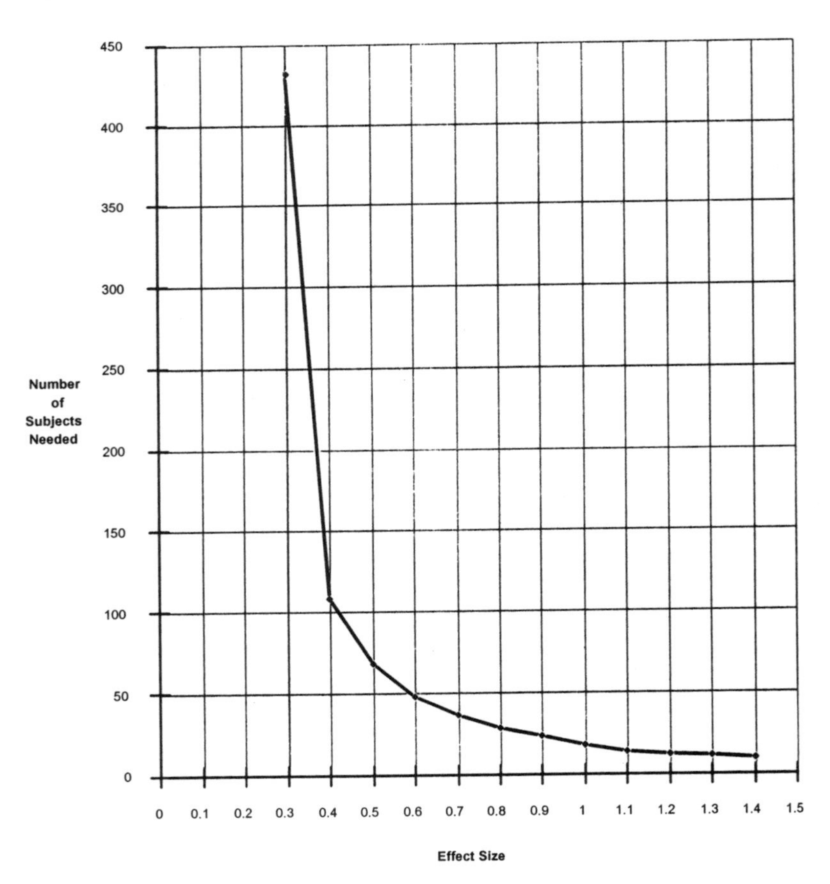

FIG. 1. Number of subjects needed as a function of effect size.

Step 8: Match Trials

Match the experimental trials to the end usage. As in Step 4, if you want to generalize the results of your experiment to what will happen in the real world, you must match the experimental trials to the real world. (Note, a single trial is defined as continuous data collection under the same experimental conditions. For example, three successive instrument approaches with same flight-control configuration constitute one trial.) The following are important characteristics to match.

Characteristic 1: Length of the Trial. Over the length of a trial, performance improves due to warm-up effects and learning and then degrades as fatigue sets in. If you measure performance in the experiment for 10 minutes but in the real world, pilots and navigators perform the task for 2 hours, your results may not reflect the peak warm up nor the peak fatigue. Consequently, you may give the lead engineer the wrong answer. So always try to match the length of each experimental trial to the length of the task in the real world.

Characteristic 2: Level of Difficulty. If you make the experimental task too easy, all the subjects will get the same performance score: 0 errors. If all the performance scores are the same, you will not be able to distinguish between experimental conditions. To avoid this problem, make the task realistically difficult. In general, the more difficult the task in the experiment, the more likely you are to find a statistical difference between experimental conditions. This is because difficulty enhances discriminability between experimental conditions. However, there are two exceptions that should be avoided in any experiment. First, if the experimental task is too difficult, the performance of all the subjects will be exactly the same: 100% errors. You will have no way of knowing which experimental condition is better and the experiment was useless. Second, if you increase the difficulty of the task beyond that which can ever be expected in the real world, you may have biased your results. In our example, you may have found that track-up displays are better than north-up displays in mountainous terrain, flying under 100 feet Above Ground Level (AGL) at speeds exceeding 500 knots with wind gusts over 60 knots. But how are they in hilly terrain, flying at 1000 feet AGL at 200 knots with wind gusts between 10 and 20 knots that is, in the conditions in which they will be used nearly 70% of the time? You can not answer this question from the results of your experiment - or if you give an answer, it may be incorrect. Therefore, these conditions should be conditions of the experiment.

Characteristic 3: Environmental Conditions. Just as in Step 4 where you tried to match the subjects to the end users, you should also try to match the environmental conditions of the laboratory (even if that laboratory is an operational aircraft or an in-flight simulator) to the environmental conditions of the real world. This is extremely important because environmental conditions can have a greater effect on performance than the independent variables in your experiment. Important environmental conditions that should be matched include lighting, temperature, noise, and task load. Lighting conditions should be matched in luminance level (possible acuity differences), position of the light source (possible glare), and type of light source (incandescent lights have "hot spots" that can create point sources of glare; fluorescent lights provide even, moderate light levels; sunlight can mask some colors and create large glare spots). Temperatures above 80° Fahrenheit decrease the amount of effort subjects expend; temperatures below 30° Fahrenheit make fine motor movements (e.g., setting radio frequencies) difficult. Noise can either enhance or degrade performance: enhancements are due to increased attention; degradations are due to distractions. Meaningful noise (e.g., a conversation) is especially distracting. Task load refers to both the number and types of tasks that are being performed at the same time as your experimental task. In general, the greater the number of tasks that are being performed simultaneously and the greater the similarity of the tasks that are being performed simultaneously, the worse the performance on the experimental task. The classic example is monitoring three radio channels simultaneously. If the volume or quality of the communications is not varied (thus making the tasks less similar), this task is extremely difficult.

Step 9: Select Data-Recording Equipment

In general, the data-recording equipment should be able to record data for 1.5 times the length of the experimental trial. This allows for false starts without changing the data tape, disk, or other storage medium. The equipment should be able to have separate channels for each continuous dependent variable (e.g., altitude, airspeed) and as many channels as necessary to record the discrete variables (e.g., reaction time (RT) to a simulated fire) without any possibility of recording the discrete variables simultaneously on the same channel (thus losing valuable data).

Step 10: Decide Subject Participation

Decide if each subject should participate in all levels. There are many advantages of having a single subject participate in more than one experimental condition: (a) reduced recruitment costs, (b) decreased total training time, and (c) better matching of subjects across experimental conditions. But there are some conditions that preclude using the same subject in more than one experimental condition. The first is previous training. In our example, pilots and navigators have had very different training. The differences in their training may affect their performance; therefore, they cannot participate in both roles: pilot and navigator. Second, some experimental conditions can make the subjects' performance worse than even untrained subjects in another experimental condition. This effect is called negative transfer. Negative transfer is especially strong when two experimental conditions require a subject to give a different response to the same stimulus. For example, the response to a fire alarm in Experimental Condition 1 is pull the T handle, then feather the engine. In Experimental Condition 2, the response is feather the engine and then pull the T handle. Subjects who have not participated in any experimental condition are going to have faster RTs and fewer errors than subjects who have already participated in either Experimental Condition 1 or 2. Whenever there is negative transfer (easy to find by comparing performance of new subjects to subjects who have already participated in another condition), use separate subjects.

Learning is another important condition affecting the decision to use the same subjects or not. Subjects who participate in more than one experimental condition are constantly learning about the task that they are performing. If you plot the subject's performance (where high scores mean good performance) on the ordinate and the number of trials he has completed along the abscissa, you will find a resulting J curve where a lot of improvement in performance occurs in the first few trials and very little improvement occurs in the final trials. The point at which there is very little improvement is called *asymptotic learning*. Unless subjects are all trained to asymptote before the first trial, their performance will improve over the entire experiment regardless of the differences in the experimental conditions. Therefore, the "improvement" you see in later experimental conditions may have nothing to do with what the experimental condition is but rather with how long the subject has been performing the task in the entire experiment.

A similar effect occurs in simple, repetitive, mental tasks and all physically demanding tasks. This effect is called *warm-up*. If the subject's performance improves over trials regardless of the experimental conditions, you may have a warm-up effect. This effect can be eliminated by having subjects perform preliminary trials until their performance on the task matches their asymptotic learning.

The final condition is fatigue. If the same subject is performing more than one trial, fatigue effects may begin to mask the differences in the experimental conditions. You can check for fatigue effects in four ways: by performing a number of trials yourself (how are you feeling?); by observing your subjects (are they showing signs of fatigue?); by comparing performance in the same trial number but different conditions across subjects (is everyone doing poorly after three trials?); and by asking the subjects how they are feeling.

Step 11: Order the Trials

In Step 10, we described order or carry over effects. Even if these do not occur to a great degree or if they do not seem to occur at all, it is still important to order your data-collection trials so as to minimize order and carry over effects. Another important carry over effect is the experimenter's experience–during your first trial, experimental procedures may not yet be smoothed out. By the 10th trial, everything should be running efficiently and you may even be anticipating subjects' questions before they ask them. The best way to minimize order and carry over effects is to use a Latin-square design. This design ensures that every experimental condition precedes and succeeds every other experimental condition an equal number of times.

Once the Latin square is generated, check the order for any safety constraints (e.g., landing a Level 3 aircraft in maximum turbulence or severe crosswinds). Adjust this order as necessary to maintain safety. The resulting numbers indicate the order in which you should collect your data. For example, Subject 1 gets north up then track up. Subject 2 gets the opposite. Once you have completed data collection for the pilots, you can collect data on the navigators. It does not matter what order you collect the pilots and navigators data because the pilots' data will never be compared to the navigators' data that is, you are not looking for an interaction between the two independent variables. If the second independent variable in the experiment had been size (e.g., the lead engineer gives you the option for an 8- or 12-inch display), the interaction would have been of interest. For example, are 12-inch, track-up displays better than 8-inch, north-up displays? If we had been interested in this interaction, a Latin square for four conditions: Condition 1, 8-inch, north up; Condition 2, 8-inch, track up; Condition 3, 12-inch, north up; and Condition 4, 12-inch, track up would have been used.

Step 12: Check for Range Effects

Range effects occur when your results differ based on the range of experimental conditions that you use. For example, in Experiment 1 you compare track-up and north-up displays, and find that for pilots track-up displays are better. In Experiment 2, you compare track-up, north-up, and horizontal situation indicator (HSI) displays. This time you find no difference between track-up and north-up displays but both are better than a conventional HSI. This is an example of a range effect: when you compare across one range of conditions, you get one answer; when you compare across a second range of conditions, you get another answer. Range effects are especially prevalent when you vary environmental conditions such as noise level and temperature. Range effects cannot be eliminated. This makes selecting a range of conditions for your experiment especially important.

To select a range of conditions, first return to your original question. If the lead engineer is asking which of two displays to use, Experiment 1 is the right experiment. If he is asking whether track-up or north-up displays are better than an HSI, Experiment 2 is correct. Second, you have to consider how many experimental conditions your subjects are experiencing. If it is more than seven, your subject is going to have a hard time remembering what each condition was but his performance will still show the effect. To check for a "number of trials" effect, plot the average performance in each trial versus the number of the trials the subject has completed. If you find a general decrease in performance, it is time to either reduce the number of experimental conditions which the subject experiences or provide long rest periods.

1.2 Summary

The quality and validity of the data are improved by incorporating the following steps in the experimental design:

Step 1: Clearly define the question to be answered.

Step 2: Check for qualifiers.

Step 3: Specify the exact conditions to be compared.

Step 4: Match the subjects to the end users.

Step 5: Select performance measures.

Step 6: Use enough subjects.

Step 7: Select data-collection equipment.

Step 8: Match the experimental trials to the end usage.

Step 9: Select data-recording equipment.

Step 10: Decide if each subject should participate in all levels.

Step 11: Order the trials.

Step 12: Check for range effects.

Step 5 is the focus for the remainder of this book.

1.3 References

Baird, J.C. and Noma, E. Fundamentals of scaling and psychophysics. New York: Wiley, 1978.

Keppel, G. Design and Analysis A Researcher's Handbook. Englewood Cliffs, NJ: Prentice Hall, 1991.

Kirk, R.R. Experimental design: Procedures for the behavioral sciences. Pacific Grove, CA: Brooks/Cole Publishing Company, 1995.

Meister, D. Human factors testing and evaluation. New York: Elsevier, 1986.

Morrow, J.R., Jackson, A.W., Disch, J.G., and Mood, D.P. Measurement and evaluation in human performance. Champaign, IL: Human Kinematics, 1995.

Torgerson, W.S. Theory and methods of scaling. New York: Wiley, 1958.

Young, F.W. Scaling. Annual Review of Psychology. 35: 55-81, 1984.

2 Human Performance

Human performance is the accomplishment of a task by a human operator. Tasks can vary from the simple (card sorting) to the complex (landing an aircraft). Humans can perform the task manually or monitor an automated system. In every case, human performance can be measured and this handbook can help. It has been highly formatted to provide the information needed to select measures that fit the criteria listed in chapter 1. For uniformity and ease of use, each discussion of a measure of human performance has the same sections (see italized section titles below):

General description of the measure;

Strengths and limitations or restrictions of the measure, including any known proprietary rights or restrictions as well as validity and reliability data;

Data collection, reduction, and analysis requirements;

Thresholds, the critical levels of performance above or below that the researcher should pay particular attention; and

Sources of further information and references.

2.1 Accuracy

General description - Accuracy is a measure of the quality of a behavior. Measures of accuracy include number correct, percent correct, and probability of correct detections. Error can also be used to measure accuracy—or the lack thereof. Error measures include absolute error, average range scores, deviations, error rate, false alarm rate, number of errors, percent errors, and root mean square error. Errors can be of omission (i.e., leaving a task out) or commission (i.e., doing the task but not correctly). Sometimes accuracy and error measures are combined to provide ratios.

Strengths and limitations - Accuracy can be measured on a ratio scale and is, thus, mathematically robust. However, distributions of the number of errors or the number correct may be skewed and, thus, may require mathematical transformation into a normal distribution. In addition, some errors rarely occur and are, therefore, difficult to investigate (Meister, 1986).

2.1.1 Absolute Error

Mertens and Collins (1986) used absolute and root mean square error on a two-dimensional compensatory tracking task to evaluate the effects of age (30 to 39 versus 60 to 69 years old), sleep (permitted versus deprived) and altitude (ground versus 3810 m). Performance was not significantly affected by age but was significantly degraded by sleep deprivation and altitude. Similar results occurred for a problem-solving task.

Elvers, Adapathya, Klauer, Kancler, and Dolan (1993) reported that as the probability of a task requiring the subject to determine a volume rather than a distance increased, absolute error of the distance judgment increased.

2.1.2 Average Range Score

Rosenberg and Martin (1988) used average range scores ("largest coordinate value minus smallest coordinate value" p. 233) to evaluate a digitizer puck (i.e., a cursor-positioning device for digitizing images). There was no effect of type of optical sight, however, magnification improved performance.

2.1.3 Deviations

Ash and Holding (1990) used timing accuracy (i.e., difference between the mean spacing between notes played and the metronome interval) to evaluate keyboard-training methods. There was a significant training method effect but no significant trial or order effects on this measure.

Yeh and Silverstein (1992) asked subjects to make spatial judgments of simplified aircraft landing scenes. They reported that subjects were less accurate in making altitude judgments relative to depth judgments. However, altitude judgments (mean percent correct) were more accurate as altitude increased and with the addition of binocular disparity.

2.1.4 Error Rate

Wierwille, Rahimi, and Casali (1985) reported that error rate was significantly related to the difficulty of a mathematical problem-solving task. Error rate was defined as the number of incorrectly answered and unanswered problems divided by the total number of problems presented.

Error rates were not significantly different among five feedback conditions (normal, auditory, color, tactile, and combined) for computer mouse use (Akamatsu, MacKenzie, and Hasbroucq, 1995). Error rates were higher for rapid communication than for conventional visual displays (Payne and Lang, 1991).

2.1.5 False Alarm Rate

Lanzetta, Dember, Warm, and Berch (1987) reported that the false alarm rate significantly decreased as the presentation rate increased (9.5%, 6/minute; 8.2%, 12/minute; 3.2%, 24/minute; 1.5%, 48/minute).

Loeb, Noonan, Ash, and Holding (1987) found the number of false alarms was significantly lower in a simple than a complex task.

Galinsky, Warm, Dember, Weiler, and Scerbo (1990) used false alarm rate to evaluate periods of watch (i.e., five 10-minute intervals). False alarm rate significantly increased as event rate decreased (5 versus 40 events per minute).

2.1.6 Number Correct

Craig, Davies, and Matthews (1987) reported a significant decrease in the number of correct detections as event rate increased, signal frequency decreased, time on task increased, and the stimulus degraded.

Loeb, Noonan, Ash, and Holding (1987) found no significant effects of cueing, constant or changing target, or brief or persistent target on number correct.

Tzelgov, Henik, Dinstein, and Rabany (1990) used the number correct to evaluate two types of stereo picture compression. This measure was significantly different between tasks (higher in object decision task than in-depth decision task), depth differences (greater at larger depth differences), and presentation condition. There were also significant interactions.

Van Orden, Benoit, and Osga (1996) used average number of correct to evaluate the effect of cold stress on a command and control task. There were no significant differences.

2.1.7 Number of Errors

Vermeulen (1987) used the number of errors to evaluate presentation modes for a system-state identification task and a process-control task. There was no mode effect on errors in the first task; however, inexperienced personnel made significantly more errors than experienced personnel. For the second task, the functional presentation mode was associated with fewer errors than the topographical presentation. Again, inexperienced personnel made more errors than experienced personnel.

Downing and Sanders (1987) reported significantly more errors were made in a simulated control room emergency with a mirror than with a nonmirror image control panel layout.

Casali, Williges, and Dryden (1990) used the number of uncorrected errors to evaluate a speech recognition system. There were significant recognition accuracy and available vocabulary effects but not a significant age effect.

Chapanis (1990) used the number of errors to evaluate short-term memory for numbers. He reported large individual differences (71 to 2231 errors out of 8,000 numbers). He also reported a significant serial position effect (70% of the subjects made the greatest number of errors at the seventh position). Women made significantly more errors than men did.

Frankish and Noyes (1990) used the number of data entry errors to evaluate four types of feedback: 1) concurrent visual feedback, 2) concurrent spoken feedback, 3) terminal visual feedback, and 4) terminal spoken feedback. There were no significant differences.

Van Orden, Benoit, and Osga (1996) used the average number of incorrect responses to evaluate the effect of cold stress on a command and control task. There were no significant differences.

2.1.8 Percent Correct

Lanzetta, Dember, Warm, and Berch (1987) reported a significantly higher percentage of correct detections in a simultaneous (78%) than in a successive (69%) vigilance task. In addition, the percentage correct generally decreased as the presentation rate increased (79%, 6/minute; 82%, 12/minute; 73%, 23/minute; 61%, 48/minute). This difference was significant. Percentage correct also significantly decreased as a function of time on watch (87%, 10 minutes; 75%, 20 minutes; 68%, 30 minutes; 65%, 40 minutes).

Chen and Tsoi (1988) used the percent of correct responses to comprehension questions to evaluate readability of computer displays. Performance was better in the slow (100 words per minute) than in the fast (200 words per minute) condition. It was also better when there were jumps of 1 rather than 5 or 9 character spaces. But there was no significant difference between 20- or 40-character windows. However, Chen, Chan, and Tsoi (1988) reported *no* significant effect of jump length but a significant effect of window size. Specifically, there was a higher comprehension score for the 20-character than for the 40-character window. The significant interaction indicated that this advantage only occurred in the one-jump condition.

Chong and Triggs (1989) used the percent of correct responses to evaluate the effects of type of windscreen post on target detections. There were significantly smaller percentages of correct detections for solid or no posts than for open posts.

Coury, Boulette, and Smith (1989) reported that percent correct was significantly greater for bargraph displays than for digital or configural displays after extended practice (8 trial blocks).

Doll and Hanna (1989) forced subjects to maintain a constant percent of correct responses during a detection task.

Imbeau, Wierwille, Wolf, and Chun (1989) used the percent of correct answers to evaluate instrument panel lighting in automobiles. They reported that accuracy decreased as character size decreased.

Matthews, Lovasik, and Mertins (1989) reported a significantly lower percent or correct responses for red on green displays (79.1%) and red on blue displays (75.5%) than for monochromatic (green, red, or blue on black), achromatic (white on black), or blue on green displays (85.5%). In addition, performance was significantly worse on the first (79.8%) and last (81.8%) half hours than for the middle three hours (83.9%).

Arnaut and Greenstein (1990) reported no significant difference in the percentage of control responses resulting in errors as a function of level of control input.

Galinsky, Warm, Dember, Weiler, and Scerbo (1990) used percentage of correct detections to evaluate periods of watch (i.e., five 10-minute intervals). This percentage decreased as event rate increased from 5 to 40 events per minute in two (continuous auditory and continuous visual) conditions but not in a third (sensory alternation) condition.

Fisk and Hodge (1992) reported a significant decrease in percent correct in a visual search task after 30 days in only one of five groups (the same category and examplars were used in training). There were no differences for new, highly related, moderately related, or unrelated examplars.

Adelman, Cohen, Bresnick, Chinnis, and Laskey (1993) used the mean percent of correct responses to evaluate operator performance with varying types of expert systems. The task was simulated in-flight communication. The experts were of three types: 1) with rule-generation capability, 2) without rule-generation capability, and 3) totally automated. The operator could screen, override, or provide a manual response. Rule-generation resulted in significantly better performance than no rule generating capability.

Lee and Fisk (1993) reported extremely small changes (1 to 290) in percent correct as a function of the consistency in targets in a visual search task.

Brand and Judd (1993) reported a significantly lower percent of correct responses for keyboard entry as the angle of the hardcopy of which they were to enter by keyboard increased (89.8% for 90 degree, 91.0% for 30 degree, and 92.4% for 12 degree). Experienced users had significantly higher percent of correct responses (94%) than naive users (88%).

Kimchi, Gopher, Rubin, and Raij (1993) reported a significantly higher percent of correct responses on a locally-directed than a globally-directed task in a divided attention condition. There were no significant attention effects on percent correct on the globally-directed task.

Donderi (1994) used percentage of targets detected to evaluate types of search for life rafts at sea. The daytime percent of correct detections were positively correlated with low contrast visual acuity and negatively correlated with vision test scores.

Briggs and Goldberg (1995) used percent of correct recognition of armored tanks to evaluate training. There were significant differences in presentation time (longer times were associated with higher accuracies), view (flank views were more accurate than frontal views), model (M1 had the highest accuracy, British Challenger had the worst accuracy), and subjects. There were no significant effects of component shown or friend versus foe.

2.1.9 Percent Errors

Hancock and Caird (1993) reported significant increases in the percent of errors as the shrink rate of a target decreased. The greatest percent of errors occurred for paths with 4 steps rather than 2, 8, or 16 steps.

2.1.10 Probability of Correct Detections

The probability of correct detections of target words was significantly higher for natural speech than for synthetic speech (Ralston, Pisoni, Lively, Greene, and Mullennix, 1991). In a similar measure, Zaitzeff (1969) reported greater cumulative target acquisition probability for two- than for one-man crews.

2.1.11 Ratio of Number Correct/Number Errors

Ash and Holding (1990) used number of errors divided by number of correct responses to evaluate keyboard-training methods. They reported significant learning between the first and third trial blocks, significant differences between training methods, and a significant order effect.

2.1.12 Root Mean Square Error

Eberts (1987) used root mean square error (rmse) on a second-order, compensatory tracking task to evaluate the effects of cueing. Cueing significantly decreased error. Vidulich (1991) reported test-retest reliability of rmse on a continuous tracking task of +0.945. The reliability of percent correct was +0.218.

Data requirements - All correct answers must be identified prior to the start of the experiment. Errors should be reviewed to ensure that they are indeed errors and not alternative versions of the correct answers.

Thresholds - During data reduction, negative numbers of errors or of correct answers should be tested for accuracy. Percent correct or percent errors greater than 100% should also be tested for accuracy.

Sources -

Adelman, L., Cohen, M.S., Bresnick, T. A., Chinnis, J.O., and Laskey, K.B. Real-time expert system interfaces, cognitive processes, and task performance: An empirical assessment. Human Factors. 35(2), 243-261; 1993.

Akamatsu, M., MacKenzie, I.S., and Hasbroucq, T. A comparison of tactile, auditory, and visual feedback in a pointing task using a mouse-type device. Ergonomics. 38(4), 816-827, 1995.

Arnaut, L.Y. and Greenstein, T.S. Is display/control gain a useful metric for optimizing an interface? Human Factors. 32(6), 651-663; 1990.

Ash, D.W. and Holding, D.H. Backward versus forward chaining in the acquisition of a keyboard skill. Human Factors. 32(2); 139-146; 1990.

Brand, J.L. and Judd, K.W. Angle of hard copy and text-editing perfomormance. Human Factors. 35(1), 57-70; 1993.

Briggs, R.W. and Goldberg, J.H. Battlefield recognition of armored vehicles. Human Factors. 37(3); 596-610; 1995.

Casali, S.P., Williges, B.H., and Dryden, R.D. Effects of recognition accuracy and vocabulary size of a speech recognition system on task performance and user acceptance. Human Factors. 32(2); 183-196; 1990.

Chapanis, A. Short-term memory for numbers. Human Factors. 32(2); 123-137; 1990.

Chen, H., Chan, K., and Tsoi, K. Reading self-paced moving text on a computer display. Human Factors. 30(3), 285-291; 1988.

Chen, H. and Tsoi, K. Factors affecting the readability of moving text on a computer display. Human Factors. 30(1); 25-33; 1988.

Chong, J. and Triggs, T.J. Visual accommodation and target detection in the vicinity of a window post. Human Factors. 31(1), 63-75; 1989.

Coury, B.G., Boulette, M.D., and Smith, R.A. Effect of uncertainty and diagnosticity on classification of multidimensional data with integral and separable displays of system status. Human Factors. 31(5), 551-569; 1989.

Craig, A., Davies, D.R., and Matthews, G. Diurnal variation, task characteristics, and vigilance performance. Human Factors. 29(6), 675-684; 1987.

Doll, T.J. and Hanna, T.E. Enhanced detection with bimodal sonar displays. Human Factors. 31(5): 539-550; 1989.

Donderi, D.C. Visual acuity, color vision, and visual search performance. Human Factors. 36(1), 129-144; 1994.

Downing, J.V. and Saunders, M.S. The effects of panel arrangement and focus of attention on performance. Human Factors. 29(5), 551-562; 1987.

Eberts, R. Internal models, tracking strategies, and dual-task performance. Human Factors. 29(4), 407-420; 1987.

Elvers, G.C., Adapathya, R.S., Klauer, K.M., Kancler, D.E., and Dolan, N.J. Effects of task probability on integral and separable task performance. Human Factors. 35(4), 629-637, 1993.

Fisk, A.D. and Hodge, K.A. Retention of trained performance in consistent mapping search after extended delay. Human Factors. 34(2), 147-164; 1992.

Frankish, C. and Noyes, J. Sources of human error in data entry tasks using speech input. Human Factors. 32(6); 697-716; 1990.

Galinsky, T.L., Warm, J.S., Dember, W.N., Weiler, E.M., and Scerbo, M.W. Sensory alternation and vigilance performance: the role of pathway inhibition. Human Factors. 32(6); 717-728; 1990.

Hancock, P.A. and Caird, J.K. Experimental evaluation of a model of mental workload. Human Factors. 35(3), 413-319; 1993.

Imbeau, D., Wierwille, W.W., Wolf, L.D., and Chun, G.A. Effects of instrument panel luminance and chromaticity on reading performance and preference in simulated driving. Human Factors. 31(2): 147-160; 1989.

Kimchi, R., Gopher, D., Rubin, Y., and Raij, D. Performance under dichoptic versus binocular viewing conditions: effects of attention and task requirements. Human Factors. 35(1), 35-56; 1993.

Lanzetta, T.M., Dember, W.N., Warm, J.S., and Berch, D.B. Effects of task type and simulus heterogeneity on the event rate function in sustained attention. Human Factors. 29(6), 625-633; 1987.

Lee, M.D. and Fisk, A.D. Disruption and maintenance of skilled visual search as a function of degree of consistency. Human Factors. 35(2), 205-220; 1993.

Loeb, M., Noonan, T.K., Ash, D.W., and Holding, D.H. Limitations of the cognitive vigilance decrement. Human Factors. 29(6), 661-674; 1987.

Matthews, M.L., Lovasik, J.V., and Mertins, K. Visual performance and subjective discomfort in prolonged viewing of chromatic displays. Human Factors. 31(3): 259-271; 1989.

Meister, D. Human factors testing and evaluation. New York: Elsevier; 1986.

Mertens, H.W. and Collins, W.E. The effects of age, sleep deprivation, and altitude on complex performance. Human Factors. 28(5): 541-551; 1986.

Payne, D.G. and Lang, V.A. Visual monitoring with spatially versus temporally distributed displays. Human Factors. 33(4), 443-458; 1991.

Ralston, J.V., Pisoni, D.B., Lively, S.E., Greene, G.B., and Mullennix, J.W. Comprehension of synthetic speech produced by rule: word monitoring and sentence-by-sentence listening tones. Human Factors. 33(4), 471-491; 1991.

Rosenberg, D.J. and Martin, G. Human performance evaluation of digitizer pucks for computer input of spatial information. Human Factors. 30(2), 231-235; 1988.
Tzelgov, J., Henik, A., Dinstein, I., and Rabany, J. Performance consequences of two types of stereo picture compression. Human Factors. 32(2); 173-182; 1990.
Van Orden, K.F., Benoit, S.L., and Osga, G.A. Effects of cold air stress on the performance of a command and control task. Human Factors. 38(1); 130-141, 1996.
Vermeulen, J. Effects of functionally or topographically presented process schemes on operator performance. Human Factors. 29(4); 383-394; 1987.
Vidulich, M.A. The Bedford Scale: Does it measure spare capacity. Proceedings of the 6th International Symposium on Aviation Psychology. 2, 1136-1141; 1991.
Wierwille, W.W., Rahimi, M., and Casali, J.G. Evaluation of 16 measures of mental workload using a simulated flight task emphasizing mediational activity. Human Factors. 27(5), 489-502; 1985.
Yeh, Y. and Silverstein, L. D. Spatial judgments with monoscopic and stereoscopic presentation of perspective displays. Human Factors. 34(5), 583-600, 1992.
Zaitzeff, L.P. Aircrew task loading in the boeing multimission simulator. Proceedings of Measurement of Aircrew Performance - The Flight Deck Workload and its Relation to Pilot Performance (AGARD-CP-56). Neuilly-sur-Seine, France: AGARD; 1969.

2.2 AGARD's Standardized Tests for Research with Environmental Stressors (STRES) Battery

General description - The STRES Battery is made up of seven tests:

1. reaction time,
2. mathematical processing,
3. memory search,
4. spatial processing,
5. unstable tracking,
6. grammatical reasoning, and
7. dual task performance of Tests 3 and 5.

Strengths and limitations - Tests were selected for the STRES Battery based on the following criteria: "(1) preliminary evidence of reliability, validity, and sensitivity, (2) documented history of application to assessment of a range of stressor effects, (3) short duration (maximum of three minutes per trial block), (4) language-independence, (5) sound basis in [Human Performance Theory] HPT, [and] (6) ability to be implemented on simple and easily-available computer systems" (AGARD, 1989, p. 7).

Data requirements - Each STRES Battery test has been programmed for computer administration. The order of presentation is fixed, as presented above. Standardized instructions are used as well as a standardized data file format. Test stimuli must be presented in white on a dark background. The joystick must have 30-degree lateral travel from the vertical position, friction not greater than 50g, linear relationship between angular rotation and lateral movement, and 8-bit resolution.

Thresholds - Not stated.

Sources -

AGARD. Human performance assessment methods (AGARD-AG-308). Neuilly-sur-Seine, France, AGARD; June 1989.

2.3 Aircraft Parameters

General description - Aircrew performance is often estimated from parameters describing aircraft state. These parameters include air speed, altitude, bank angle, descent rate, glideslope, localizer, pitch rate, roll rate, and yaw rate. Measures derived from these include root-mean-square values, minimums and maximums, correlations between two or more of these parameters, and deviations between actual and assigned values.

Strengths and limitations – Aircraft parameters are sensitive to phase of flight (takeoff, climb, cruise, approach, landing) or flight mode (hover). Hardy and Parasuraman (1997) developed an error list based on component pilot activities (see Table 1).

TABLE 1.
Component Abilities of Commercial Airline Pilot Performance Determined by Frequency of Errors Extracted from Accident Reports, Critical Incidents, and Flight Checks

Component Ability	Frequency of Errors			
	Accidents	Incidents	Flight Checks	Total
Establishing and maintaining angle of glide, rate of descent, and gliding speed on approach to landing	47	41	11	99
Operating controls and switches	15	44	33	92
Navigating and orienting	4	39	19	62
Maintaining safe airspeed and attitude, recovering from stalls and spins	11	28	18	57
Following instrument flight procedures and observing instrument flight regulations	5	27	13	45
Carrying out cockpit procedures and routines	7	31	4	42
Establishing and maintaining alignment with runway on approach or takeoff climb	3	31	5	39
Attending, remaining alert, maintaining lookout	14	23	1	38
Utilizing and applying essential pilot information	0	19	18	37
Reading, checking, and observing instruments, dials, and gauges	1	26	7	34
Preparing and planning of flight	2	27	3	32
Judging type of landing or recovering from missed or poor landing	1	23	8	32
Breaking angle of glide on landing	1	25	5	31
Obtaining and utilizing instructions and information from control personnel	3	21	0	24
Reacting in an organized manner to unusual or emergency situations	0	17	7	24
Operating plane safely on ground	7	15	1	23
Flying with precision and accuracy	0	7	15	22
Operating and attending to radio	0	7	10	17
Handling of controls smoothly and with coordination	0	6	8	14
Preventing plane from undue stress	0	5	7	12
Taking safety precautions	2	5	4	11

Note: Based on "The Airline Pilot's Job," (Gordon, 1949).

Hardy and Parasuraman, 1997

2.3.1 *Takeoff and Climb*

Wierwille, Rahimi, and Casali (1985) reported that neither pitch nor roll high-pass mean square scores from a primary simulated flight task were sensitive to changes in the difficulty of a secondary mathematical problem-solving task.

Kelly (1988) reviewed approaches to automated aircrew performance measurement. He concluded that measures must include positional advantage or disadvantage, control manipulation, and management of kinetic and potential energy.

2.3.2 *Cruise*

In an in-flight helicopter study, Berger (1977) reported significant differences in air speed, altitude, bank angle, descent rate, glideslope, localizer, pitch rate, roll rate, and yaw rate and some combination of Visual Meteorological Conditions (VMC), Instrument Meteorological Conditions (IMC) with fixed sensor, IMC with stabilization sensor, and IMC with a sensor looking ahead through turns.

North, Stackhouse, and Graffunder (1979) reported that: (1) pitch error, heading error, roll acceleration, pitch acceleration, speed error, and yaw position were sensitive to differences in display configurations, (2) pitch error, roll error, heading error, roll acceleration, pitch acceleration, yaw acceleration, speed error, pitch position, roll position, yaw position, power setting, altitude error, and crosstrack error were sensitive to differences in winds, and (3) heading error, roll acceleration, pitch acceleration, yaw acceleration, speed error, roll position, yaw position, altitude error, and crosstrack error were sensitive to motion cues.

Kruk, Regan, and Beverley (1983) compared the performance of 12 experienced fighter pilots, 12 undergraduate training instructor pilots, and 12 student pilots on three tasks (formation, low-level, and landing) in a ground simulator. Only one parameter was sensitive to flight experience - students were significantly poorer than instructors in the distance of first correction to the runway during the landing task. The measures were: time spent in correct formation, percentage of bomb strikes within 36m of target center, time in missile tracking, number of times shot down, number of crashes, altitude variability, heading variability, release height variability, and gravity load at release.

Bortolussi and Vidulich (1991) calculated a figure of merit (FOM) for simulated flight from rudder standard deviation (SD), elevator SD, aileron SD, altitude SD, mean altitude, airspeed SD, mean airspeed, heading SD, and mean heading. Only the airspeed and altitude FOMs showed significant differences between scenarios. The primary measures for these variables were also significant as well as aileron SD and elevator SD.

Barfield, Rosenberg, and Furness (1995) examined the effect of frame of reference, field of view (FOV), and eyepoint elevation on performance of a simulated air-to-ground targeting task. There was a significant frame of reference effect. Specifically, the pilot's eye display was associated with lower flight-path rmse; faster target lock-on time, and faster target acquisition time than the God's eye display. There was also a significant effect of FOV: lower rmse for 30° and 90° FOV than for 60° FOV, fastest time to lockout target for 30° FOV, and fastest target acquisition time. Finally, eyepoint elevation also resulted in significant differences: lower rmse, faster lock on times, and faster target acquisition times for 60° than 30° elevation.

2.3.3 Approach and Landing

Brictson (1969) reported differences between night and day carrier landings for the following aircraft parameters: altitude error from glideslope, bolster rate, wire arrestments, percent of unsuccessful approaches, and probability of successful recovery. There were no significant day–night differences for lateral error from centerline, sink speed, and final approach airspeed.

Kraft and Elworth (1969) reported significant differences in generated altitude during final approach due to individual differences, city slope, and lighting.

Swaroop and Ashworth (1978) reported significantly higher glideslope intercepts and flight path elevation angles with than without a diamond marking on the runway. Touchdown distance was significantly smaller without the diamond for research pilots and with the diamond for general aviation pilots.

Lintern, Kaul, and Collyer (1984) reported significant differences between conventional and modified Fresnel Lens Optical Landing System displays for glideslope rmse and descent rate at touchdown but not for localizer rmse.

Morello (1977), using a B-737 aircraft, reported differences in localizer, lateral, and glideslope deviations during three-nautical-mile and close-in approaches between a baseline and an integrated display format. However, neither localizer rmse nor glideslope rmse were sensitive to variations in pitch-stability level, wind-gust disturbance, or crosswind direction and velocity (Wierwille and Connor, 1983).

Gaidai and Mel'nikov (1985) developed an integral criterion to evaluate landing performance:

$$I(t) = \frac{1}{t_z} \int_0^{t_z} \sum_{j=1}^{K} a_i(t_i) \left[\frac{Y_i(t_i) - m_{yi}}{s_{yi}} \right]^2 dt$$

where

I = integral criteria

t = time

t_z = integration time

$Ka_j(t_i)$ = weighting coefficient for the path parameter Y_i at instant i

$Y_j(t_i)$ = instantaneous value of parameter Y_j at instant i

m_{Yj} = programmed value of the path parameter

s_{Yj} = standard deviation for the integral deviations:

$$s_{yj} = \frac{1}{n} \sum_{j=1}^{n} \sqrt{\frac{1}{t_z} \int_0^{t_z} [Y_j(t_i) - m_{yi}]^2 dt}$$

I(t) provides a multivariate assessment of pilot performance but is difficult to calculate.

Lintern and Koonce (1991) reported significant differences in vertical angular glideslope errors among varying scenes, display magnifications, runway sizes, and start points.

2.3.4 Hover

Moreland and Barnes (1969) derived a measure of helicopter pilot performance using the following equation:

100 - (absolute airspeed error + absolute altitude error + absolute heading error + absolute change in torque).

This measure decreased when cockpit temperature increased above 85°F, was better in light to moderate than no turbulence, and was sensitive to basic piloting techniques. However, it was not affected by either clothing or equipment configurations.

Richard and Parrish (1984) used a vector combination of errors (VCE), to estimate hover performance. VCE was calculated as follows:

$$VCE = (x^2 + y^2 + z^2)^{1/2}$$

where x, y, and z refer to the x, y, and z axis errors. The authors argue that VCE is a good summary measure since it discriminated trends in the data.

Data requirements - Not stated.

Thresholds - Not stated.

Sources -

Barfield, W., Rosenberg, C., and Furness, T.A. Situation awareness as a function of frame of reference, computer-graphics eyepoint elevation, and geometric field of view. The International Journal of Aviation Psychology. 5(3): 233-256; 1995.

Berger, I.R. Flight performance and pilot workload in helicopter flight under simulated IMC employing a forward looking sensor. Proceedings of Guidance and Control Design Considerations for Low-Altitude and Terminal-Area Flight (AGARD-CP-240). Neuilly-sur-Seine, France: AGARD; 1977.

Bortolussi, M.R. and Vidulich, M.A. An evaluation of strategic behaviors in a high fidelity simulated flight task. Comparing primary performance to a figure of merit. Proceedings of the 6th International Symposium on Aviation Psychology. 2, 1101-1106; 1991.

Brictson, C.A. Operational measures of pilot performance during final approach to carrier landing. Proceedings of Measurement of Aircrew Performance - The Flight Deck Workload and Its Relation to Pilot Performance (AGARD-CP-56). Neuilly-sur-Seine, France: AGARD; 1969.

Gaidai, B.V. and Mel'nikov, E.V. Choosing an objective criterion for piloting performance in research on pilot training on aircraft and simulators. Cybermetrics and Computing Technology. 3:162-169; 1985.

Hardy, D.J. and Parasuraman, R. Cognition and flight performance in older pilots. Journal of Experimental Psychology. 3(4): 313-348; 1997.

Kelly, M.J. Performance measurement during simulated air-to-air combat. Human Factors. 30(4):495-506; 1988.

Kraft, C.L. and Elworth, C.L. Flight deck work and night visual approach. Proceedings of Measurement of Aircrew Performance - The Flight Deck Workload and its Relation to Pilot Performance (AGARD-CP-56). Neuilly-sur-Seine, France: AGARD; 1969.

Kruk, R., Regan, D., and Beverley, K.I. Flying performance on the advanced simulator for pilot training and laboratory tests of vision. Human Factors. 25(4):457-466; 1983.

Lintern, G., Kaul, C.E., and Collyer, S.C. Glideslope descent-rate cuing to aid carrier landings. Human Factors. 26(6): 667-675; 1984.

Lintern, G. and Koonce, J.M. Display magnification for simulated landing approaches. International Journal of Aviation Psychology. 1(1): 59-72; 1991.

Moreland, S. and Barnes, J.A. Exploratory study of pilot performance during high ambient temperatures/humidity. Proceedings of Measurement of Aircrew Performance (AGARD-CP-56). Neuilly-sur-Seine, France: AGARD; 1969.

Morello, S.A. Recent flight test results using an electronic display format on the NASA B-737. Proceedings of Guidance and Control Design Considerations for Low-Altitude and Terminal Area Flight (AGARD-CP-240). Neuilly-sur-Seine, France: AGARD; 1977.

North, R.A., Stackhouse, S.P., and Graffunder, K. Performance, physiological, and oculometer evaluation of VTOL landing displays (NASA-CR-3171). Hampton, VA: NASA Langley Research Center; 1979.

Richard, G.L., and Parrish, R.V. Pilot differences and motion cuing effects on simulated helicopter hover. Human Factors. 26(3): 249-256; 1984.

Swaroop, R. and Ashworth, G.R. An analysis of flight data from aircraft landings with and without the aid of a painted diamond on the same runway (NASA-CR-143849). Edwards Air Force Base, CA: NASA Dryden Research Center; 1978.

Wierwille, W.W. and Connor, S.A. Sensitivity of twenty measures of pilot mental workload in a simulated ILS task. Proceedings of the Annual Conference on Manual Control (18th), 150-162, 1983.

Wierwille, W.W., Rahimi, M., and Casali, J.G., Evaluation of 16 measures of mental workload using a simulated flight task emphasizing mediational activity. Human Factors. 27(5), 489-502; 1985.

2.4 Armed Forces Qualification Test

General description - The Armed Forces Qualification Test measures mechanical and mathematical aptitude. Scores from the test are used to place Army recruits into a military occupational specialty.

Strengths and limitations - To minimize attrition due to poor placement, supplementary aptitude tests are required. Further, the results are affected by the substitutability with compensation principle. This principle states, "a relatively high ability in one area makes up for a low level in another so that observed performance equals that predicted from a linear combination of two predictor measures" (Uhlaner, 1972, p. 206).

Data requirements - Cognitive-noncognitive variance should be considered when evaluating the test scores.

Thresholds - Not stated.

Sources -

Uhlaner, J.E. Human performance effectiveness and the systems measurement bed. Journal of Applied Psychology. 56(3): 202-210; 1972.

2.5 Boyett and Conn's White-Collar Performance Measures

General description - Boyett and Conn (1988) identified lists of performance measures with which to evaluate personnel in white-collar, professional, knowledge-worker organizations. These lists are categorized by function being performed, i.e., engineering, production planning, purchasing, and management information systems. The lists are provided in Table 2.

Strengths and limitations - These measures are among the few developed for white-collar tasks and were developed in accordance with the following guidelines: (1) "involve white-collar employees in developing their own measures" (Boyett and Conn, 1988, p. 210), (2) "measure results, not activities" (p. 210), (3) "use group or team-based measures" (p. 211), and (4) "use a family of indicators" (p. 211).

Data requirements - Not stated.

Thresholds - Not available.

Sources -

Boyett, J.H. and Conn, H.P. Developing white-collar performance measures. National Productivity Review. Summer: 209-218; 1988.

TABLE 2.
White-Collar Measures in Various Functions

Engineering

- Percent new or in-place equipment/tooling performing as designed
- Percent machines/tooling capable of performing within established specifications
- Percent operations with current detailed process/method sheets
- Percent work run on specified tooling
- Number bills of material errors per employee
- Percent engineering change orders per drawing issued
- Percent material specification changes per specifications issued
- Percent engineering change requests to drawings issued based on design or material changes due to drawing/specification errors
- Percent documents (drawings, specifications, process sheets, etc.) issued on time

Production Planning or Scheduling

- Percent deviation actual/planned schedule
- Percent on-time shipments
- Percent utilization manufacturing facilities
- Percent overtime attributed to production scheduling
- Percent earned on assets employed
- Number, pounds, or dollars delayed orders
- Percent back orders
- Percent on-time submission master production plan
- Hours time lost waiting on materials
- Number of days receipt of work orders prior to scheduled work
- Percent turnover of parts and material (annualized)

Purchasing

- Dollar purchases made
- Percent purchases handled by purchasing department
- Dollar purchases by major type
- Percent purchases/dollar sales volume
- Percent "rush" purchases
- Percent orders exception to lowest bid
- Percent orders shipped "most economical"
- Percent orders shipped "most expeditious"
- Percent orders transportation allowance verified
- Percent orders price variance from original requisition

Purchasing (continued)

- Percent orders "cash discount" or "early payment discount"
- Percent major vendors–annual price comparison completed
- Percent purchases–corporate guidelines met
- Elapsed time–purchase to deliver
- Percent purchases under long-term or "master contract"
- Dollar adjustment obtained/dollar value "defective" or "reject"
- Purchasing costs/purchase dollars
- Purchasing costs/number purchases
- Dollar value rejects/dollar purchases
- Percent shortages

Management Information Systems

- Number of data entry/programming errors per employee
- Percent reports issued on time
- Data processing costs as percent of sales
- Number of reruns
- Total data processing cost per transaction
- Percent target dates met
- Average response time to problem reports
- Number data entry errors by type
- Percent off-peak jobs completed by 8:00 am
- Percent end-user available (prime) on-line
- Percent on-line 3-second response time
- Percent print turnaround in 1 hour or less
- Percent prime shift precision available
- Percent uninterrupted power supply available
- Percent security equipment available
- Score on user satisfaction survey
- Percent applications on time
- Percent applications on budget
- Correction costs of programming errors
- Number programming errors
- Percent time on maintenance
- Percent time on development
- Percent budget maintenance
- Percent budget development

from Boyett and Conn (1989) p. 214

2.6 Charlton's Measures of Human Performance in Space Control Systems

General description - Charlton's (1992) measures to predict human performance in space control systems are divided into three phases (pre-pass, contact execution, and contact termination) and three crew positions (ground controller, mission controller, and planner analyst). The measures by phase and crew position are: (1) pre-pass phase, ground-controller time to complete readiness tests and errors during configuration; (2) contact-execution phase, mission-controller track termination time and commanding time; (3) contact-execution phase, planner-analyst communication duration; (4) contact-termination phase, planner-analyst communication duration; and (5) contact-termination phase, ground-controller deconfiguration time, time to return resources, and time to log off system.

Strengths and limitations - The measures were evaluated in a series of three experiments using both civilian and Air Force satellite crews.

Data requirements - Questionnaires are used as well as computer-based scoring sheets.

Thresholds - Not stated.

Sources -

Charlton, S.G. Establishing human factors criteria for space control systems. Human Factors. 34: 485-501; 1992.

2.7 Control Input Activity

General description - Corwin, Sandry-Garza, Biferno, Boucek, Logan, Jonsson, and Metalis (1989) used control input activity for the wheel (aileron) and column (elevator) as a measure of flight-path control. Griffith, Gros, and Uphaus (1984) defined control reversal rate as "the total number of control reversals in each controller axis divided by the interval elapsed time" (p. 993). They computed separate control reversal rates for steady state and maneuver transition intervals in a ground-based flight simulator.

Strengths and limitations - Corwin, et al. (1989) reported that control input activity was a reliable and valid measure. Griffith, et al. (1984) were unable to compute control reversal rates for throttle and rudder pedal activity due to minimal control inputs. Further, there were no significant differences among display configurations for pitch-axis control reversal rate. There were, however, significant differences among the same display configurations in the roll-axis control reversal rate.

Wierwille, Rahimi, and Casali (1985) used the total number of elevator, aileron, and rudder inputs during simulated flight. This measure was not sensitive to difficulty of a mathematical problem-solving task performed during simulated flight.

Data requirements - Both control reversals and time must be simultaneously recorded.

Thresholds - Not available.

Sources -

Corwin, W.H., Sandry-Garza, D.L., Biferno, M.H., Boucek, G.P., Logan, A.L., Jonsson, J.E., and Metalis, S.A. Assessment of crew workload measurement methods, techniques and procedures. Volume 1 - process, methods, and results (WRDC-TR-89-7006). Wright-Patterson Air Force Base, OH; 1989.

Griffith, P.W., Gros, P.S., and Uphaus, J.A. Evaluation of pilot performance and workload as a function of input data rate and update frame rate on a dot-matrix graphics display. Proceedings of the National Aerospace and Electronics Conference, 988-995; 1984.

Wierwille, W.W., Rahimi, M., and Casali, J.G., Evaluation of 16 measures of mental workload using a simulated flight task emphasizing mediational activity. Human Factors. 27(5), 489-502; 1985.

2.8 Correctness Score

General description - Correctness scores were developed to evaluate human problem solving. A score using the following five-point rating scale was awarded based on a subject's action:

0 - Subject made an incorrect or illogical search

1 - Subject asked for information with no apparent connection to the correct response

2 - Subject asked for incorrect information based on a logical search pattern

4 -Subject was on the right track

5 - Subject asked for the key element (Giffen and Rockwell, 1984)

Strengths and limitations - Giffen and Rockwell (1984) used correctness scores to measure a subject's problem-solving performance. These authors used stepwise regression to predict correctness scores for four scenarios: "(1) an oil-pressure gauge line break, (2) a vacuum-pump failure, (3) a broken magneto drive gear, and (4) a blocked static port" (p. 575). Demographic data, experience, knowledge scores, and information-seeking behavior were only moderately related to correctness scores. The subjects were 42 pilots with 50 to 15,000 flight hours of experience.

Data requirements - Well-defined search patterns.

Thresholds - 0 (poor performance) to 5 (excellent performance).

Sources -

Griffin, W.C. and Rockwell, T.H., Computer-aided testing of pilot response to critical in-flight events. Human Factors. 26(5), 573-581; 1984.

2.9 Critical Incident Technique

General description - The Critical Incident Technique is a set of specifications for collecting data from observed behaviors. These specifications include:

(1) Persons to make the observations must have:
   (a) Knowledge concerning the activity.
   (b) Relation to those observed.
   (c) Training requirements.

(2) Groups to be observed, including:
   (a) General description.
   (b) Location.
   (c) Persons.
   (d) Times.
   (e) Conditions.

(3) Behaviors to be observed with an emphasis on:

(a) General type of activity.
(b) Specific behaviors.
(c) Criteria of relevance to general aim.
(d) Criteria of importance to general aim (critical prints).
(Flanagan, 1954, p. 339)

Strengths and limitations - The technique has been used successfully since 1947. It is extremely flexible but can be applied only to observable performance activities.

Data requirements - The specifications listed in the general description paragraph must be applied.

Thresholds - Not stated.

Sources -

Flanagan, J.C. The critical incident technique. Psychological Bulletin. 51 (4), 327-358; 1954.

2.10 Deutsch and Malmborg Measurement Instrument Matrix

General description - The measurement instrument matrix consists of activities that must be performed by an organization along the vertical axis and the metrics used to evaluate the performance of those activities along the horizontal axis. A value of one is placed in every cell in which the metric is an appropriate measure of the activity; otherwise, a zero is inserted.

Strengths and limitations - This method handles the complexity and interaction of activities performed by organizations. It has been used to assess the impact of information overload on decision-making effectiveness. The measurement instrument matrix should be analyzed in conjunction with an objective matrix. The objective matrix lists the set of activities along the horizontal axis and the set of objectives along the vertical axis.

Data requirements - Reliable and valid metrics are required for each activity performed.

Thresholds - Zero is the lower limit; one, the upper limit.

Sources -

Deutsch, S.J. and Malmborg, C.J. The design of organizational performance measures for human decision making, Part I: Description of the design methodology. IEEE Transactions on Systems, Man, and Cybernetics. SMC-12 (3), 344-352; 1982.

2.11 Dichotic Listening

General description - Subjects are presented with auditory messages through headphones. Each message contains alphanumeric words. Different messages are presented simultaneously to each ear. Subjects must detect specific messages when they are presented in a specific ear.

Strengths and limitations - Gopher (1982) reported significant differences in scores between flight cadets who completed training and those who failed.

Data requirements - Auditory tapes are required as well as an audio system capable of displaying different messages to each ear. Finally, a system is required to record omissions, intrusions, and switching errors.

Thresholds - Low threshold is zero. High threshold was not stated.

Source –

Gopher, D. selective attention test as a predictor of success in flight training. Human Factors. 24(2):173-183; 1982.

2.12 Driving Parameters

General description - Driving parameters include measures of driver behavior (e.g., average brake RT, brake pedal errors, control light response time, number of brake responses, perception-response time, speed, and steering wheel reversals) as well as measures of total system performance (e.g., time to complete a driving task, tracking error).

Strengths and limitations – Most researchers use several driving parameters in a single study. For example, Popp and Faerber (1993) used speed, lateral distance, longitudinal distance to the leading car, acceleration in all axes, steering angle, heading angle and frequency of head movements of subjects driving a straight road in a moving-base simulator to evaluate four feedback messages that a voice command was received. There were no significant differences among these dependent variables as a function of type of feedback message.

2.12.1 Average Brake RT

Drory (1985) reported significant differences in average brake RT associated with different types of secondary tasks. It was not affected by the amount of rest drivers received prior to the simulated driving task.

Morrison, Swope, and Malcomb (1986) found significant differences in movement time from the accelerator to the brake as a function of brake placement (lower than the accelerator resulted in shorter movement times) and gender (women had longer movement times except when the brake was below the accelerator).

Korteling (1990) used the RT of correct responses and error percentages to compare laboratory, stationary, and on-road driving performance. Older drivers (61 to 73 years old) and brain-injured patients had significantly longer RTs than younger drivers (21 to 43 years old). RTs were significantly longer in on-road driving than in the laboratory. There was a significant effect of Inter-Stimulus Interval (ISI). Specifically, the shortest ISI was associated with the longest RT. Patients made significantly more errors than older or younger drivers.

Korteling (1994) used four measures to assess platoon car-following performance: 1) brake RT, 2) the correlation between the speeds of the two cars, 3) time to obtain the maximum correlation, and 4) the maximum correlation. Brake RT and delay time were significantly longer for the patients than for the older or younger drivers. Brake RT was significantly longest when the driving speed of the lead car was varied and the road was winding. Both correlation measures were significantly lower for the patients and older drivers than for the younger drivers.

2.12.2 *Brake Pedal Errors*

Rogers and Wierwille (1988) used the type and frequency of pedal actuation errors to evaluate alternative brake pedal designs. They reported 297 errors over 72 hours of testing. Serious errors occurred when the wrong or both pedals were depressed. Catch errors occurred when a pedal interferred with a foot movement. If the interference was minimal, the error was categorized as a scuff. Instructional errors occurred when the subject failed to perform the task as instructed.

Vernoy and Tomerlin (1989) used pedal error, "hitting the accelerator pedal when instructed to depress the brake pedal" (p. 369), to evaluate misperception of the centerline. There was no significant difference in pedal error among eight types of automobile used in the evaluation. There was a significant difference in deviation from the centerline among the eight automobiles, however.

Szlyk, Seiple, and Viana (1995) reported a significant increase in braking response time as age increased. The data were collected in a simulator. There were no significant effects of age or visual impairment on the variability of brake pressure.

2.12.3 *Control Light Response Time*

Drory (1985) reported significant differences in control light response time associated with various types of secondary tasks. It was not affected by the amount of rest drivers received prior to the simulated driving task.

2.12.4 *Number of Brake Responses*

Drory (1985) reported significant differences in the number of brake responses used to evaluate the effects of various types of secondary tasks. The type of task did not affect the number of brake responses nor did the amount of rest drivers received prior to the simulated driving task.

2.12.5 *Perception-Response Time*

Olson and Sivak (1986) measured perception-response time from the first sighting of an obstacle until the accelerator was released and the driver contacted the brake. Their data were collected in an instrumented vehicle driven on a two-lane rural road.

2.12.6 *Speed*

Shinar and Stiebel (1986) used speed, speed above or below the speed limit, speed reduction (((original speed - speed at site 1)/original speed) x 100) and speed resumption (((speed at site 2 - speed limit)/speed limit) x 100) to evaluate the effectiveness of the presence of police cars in reducing speeding.

Szlyk, Seiple, and Viana (1995) reported a significant decrease in speed as age increased. The data were collected in a simulator.

2.12.7 *Steering Wheel Reversals*

Hicks and Wierwille (1979) reported that steering reversals were sensitive to workload (i.e., gusts at the front of a driving simulator).

Drory (1985) reported significant differences in among steering wheel reversals associated with various types of secondary tasks. It was not affected by the amount of rest drivers received prior to the simulated driving task.

Frank, Casali, and Wierwille (1988) used the number of large steering reversals (greater than 5 degrees), number of small steering reversals (less than 5 degrees), and yaw standard deviation ("angle in the horizontal plane between the simulated vehicle longitudinal axis and the instantaneous roadway tangent" p. 206) to evaluate the effects of motion system transport delay and visual system transport delay. All three measures of driver performance were significantly related to transport delay.

2.12.8 Time

Gawron, Baum, and Perel (1986) used the time to complete a double-lane change as well as the number of pylons struck to evaluate side impact padding thickness (0, 7.5, 10 cm), direction of the lane change (left/right, right/left), and replication (1 to 12). Subjects took longer to perform the left/right than the right/left lane change. Time to complete the lane change increased as padding thickness increased (0 cm = 3.175 s, 7.5 cm = 3.222 s, and 10 cm = 3.224 s). There were no significant effects on the number of pylons struck.

Sidaway, Fairweather, Sekiya, and McNitt-Gray (1996) asked subjects to estimate time to collision after viewing videotapes of accidents. Subjects consistently underestimated the time. However, as velocity increased the time estimate was more accurate.

2.12.9 Tracking Error

Hicks and Wierwille (1979) reported that yaw deviation and lateral deviation were sensitive to workload (i.e., gusts at the front of a driving simulator).

Drory (1985) reported significant differences in tracking error associated with various types of secondary tasks. It was not affected by the amount of rest drivers received prior to the simulated driving task.

Godthelp and Kappler (1988) reported larger standard deviations in lateral position when drivers were wearing safety goggles than when they were not. This measure also increased as speed increased.

Imbeau, Wierwille, Wolf, and Chun (1989) reported that if drivers failed to respond to a display reading task, the variance of lane deviation decreased. The data were collected in a driving simulator.

Korteling (1994) reported no significant differences in the standard deviation of lateral position between young (21–34) and old (65–74 years) drivers in a driving simulator. However, older drivers had significantly larger longitudinal standard deviation in a car-following task. There was also a decrement associated with fatigue in steering performance but not in car-following performance.

Szlyk, Seiple, and Viana (1995) reported a significant increase in the number of lane crossings as age increased. The data were collected in a simulator.

Summala, Nieminen, and Punto (1996) reported worse lane-keeping performance for novice drivers than for experienced drivers when the display was near the periphery rather than in the middle console of a driving simulation.

van Winsum (1996) reported that steering wheel angle increased as curve radii decreased. Steering error increased as steering wheel angle increased. The data were collected in a driving simulator.

Data requirements - An instrumented simulator or vehicle is required.

Thresholds - Total time varied between 0.1 and 1.8 seconds (Olson and Sivak, 1986).

Sources –

Drory, A. Effects of rest and secondary task on simulated truck-driving task performance. Human Factors. 27(2), 201-207; 1985.

Frank, L.H., Casali, J.G., and Wierwille, W.W. Effects of visual display and motion system delays on operator performance and uneasiness in a driving simulator. Human Factors. 30(2), 201-217; 1988.

Gawron, V.J., Baum, A.S., and Perel, M. Effects of side-impact padding on behavior performance. Human Factors. 28(6), 661-671; 1986.

Godthelp, H. and Kappler, W.D. Effects of vehicle handling characteristics on driving strategy. Human Factors. 30(2), 219-229; 1988.

Hicks, T.G. and Wierwille, W.W. Comparison of five mental workload assessment procedures in a moving-base driving simulator. Human Factors. 21(2): 129-143; 1979.

Imbeau, D., Wierwille, W.W., Wolf, L.D., and Chun, G.A. Effects of instrument panel luminance and chromaticity on reading performance and preference in simulated driving. Human Factors. 31(2): 147-160; 1989.

Korteling, J.E. Perception-response speed and driving capabilities of brain-damaged and older drivers. Human Factors. 32(1), 95-108; 1990.

Korteling, J.E. Effects of aging, skill modification, and demand alternation on multiple task performance. Human Factors. 36(1), 27-43; 1994.

Morrison, R.W., Swope, J.G., and Malcomb, C.G. Movement time and brake pedal placement. Human Factors. 28(2), 241-246; 1986.

Olson, P.L. and Sivak, M. Perception-response time to unexpected roadway hazards. Human Factors. 28(1), 91-96; 1986.

Popp, M.M. and Faerber, B. Feedback modality for nontransparent driver control actions: why not visually? In A.G. Gale, J.D. Brown, C.M. Haslegrave, H.W. Kruysse, and S.P. Taylor (Eds). Vision in vehicles - IV. Amsterdam: North-Holland; 1993.

Rogers, S.B. and Wierwille, W.W. The occurrence of accelerator and brake pedal actuation errors during simulated driving. Human Factors. 30(1), 71-81; 1988.

Shinar, D. and Stiebel, J. The effectiveness of stationary versus moving police vehicles on compliance with speed limit. Human Factors. 28(3), 365-371; 1986.

Sidaway, B., Fairweather, M., Sekiya, H., and McNitt-Gray, J. Time-to-collision estimation in a simulated driving task. Human Factors. 38(1), 101-113; 1996.

Summala, H., Nieminen, T., and Punto, M. Maintaining lane position with peripheral vision during in-vehicle tasks. Human Factors. 38(3), 442-451; 1996.

Szlyk, J.P., Seiple, W., and Viana, M. Relative effects of age and compromised vision on driving performance. Human Factors. 37(2), 430-436; 1995.

van Winsum, W. Speed choice and steering behavior in curve driving. Human Factors. 38(3), 434-441; 1996.

Vernoy, M.W. and Tomerlin, J. Pedal error and misperceived centerline in eight different automobiles. Human Factors. 31(4), 369-375; 1989.

2.13 Eastman Kodak Company Measures for Handling Tasks

General description - The Eastman Kodak Company (1986) has developed a total of eight measures to assess human performance in repetitive assembly, packing, or handling tasks. These eight measures have been divided into (1) measures of productivity over the shift: total units per shift at different levels and durations of effort and/or exposure, units per hour compared to a standard, amount of time on arbitrary work breaks or secondary work, amount of waste, and work interruptions, distractions, and accidents and (2) quality of output: missed defects/communications, improper actions, and incomplete work (p. 104).

Strengths and limitations - These measures are well suited for repetitive overt tasks but may be inappropriate for maintenance or monitoring tasks.

Data requirements - Task must require observable behavior.

Thresholds - Not stated.

Source -

Eastman Kodak Company. Ergonomic design for people at work. New York: Van Nostrand Reinhold; 1986.

2.14 Glance

General description - The duration that a human visually samples a single scene is a glance. Glance duration and location have been used to evaluate control, displays, and procedures.

Strengths and limitations - Glance duration has long been used to evaluate driver performance. In an early study, Mourant and Rockwell (1970) analyzed the glance behavior of eight drivers traveling at 50 mph on an expressway. As the route became more familiar, drivers increased glances to the right edge marker and horizon. While following a car, drivers glanced more often at lane markers.

Mourant and Donohue (1977) reported that novice and young experienced drivers made fewer glances to left outside mirror than did mature drivers. Novice drivers also made more direct looks than glances in the mirrors prior to executing a maneuver.

Imbeau, Wierwille, Wolf, and Chun (1989) used time glancing at a display to evaluate instrument panel lighting in automobiles. Not unexpectedly, higher complexity of messages was associated with significantly longer (+0.05s more) glance times.

Novice drivers have longer and more frequent glances at vehicles than at obstacles (Masuda, Nagata, Kureyama, and Sato, 1990). Glance duration increases as blood alcohol concentration level increases (Masuda, Nagata, Kureyama, and Sato, 1990).

Land (1993) reported that drivers glance behavior varies throughout a curve: a search for cues during approach, into the curve during the bend, and target beyond the bend on exit.

There has been some nondriver research as well. For example, Fukuda (1992) reported a maximum of 6 characters could be recognized in a single glance.

Data Requirements - Eye movements must be recorded to an accuracy of ±0.5° horizontal and ±1° vertical (Mourant and Rockwell, 1970).

Thresholds - Minimum glance duration 0.68s (Mourant and Donohue, 1977). Maximum glance duration 1.17s (Mourant and Donohue, 1977).

Sources -

Fukuda, T. Visual capability to receive character information Part I: how many characters can we recognize at a glance? Ergonomics. 35(5), 617-627, 1992.

Imbeau, D., Wierwille, W.W., Wolf, L.D., and Chun, G.A. Effects of instrument panel luminance and chromaticity on reading performance and preference in simulated driving. Human Factors 31(2): 147-160; 1989.

Land, M.F. Eye-head coordination during driving. IEEE Systems, Man and Cybernetics Conference Proceedings. 490-494, 1993.

Masuda, K., Nagata, M., Kureyama, H., and Sato, T.B. Visual behavior of novice drivers as affected by traffic conflicts (SAE Paper 900141). Warrendale, PA. Society of Automotive Engineers, 1990.

Mourant, R.R. and Donohue, R.J. Acquisition of indirect vision information by novice, experienced, and mature drivers. Journal of Safety Research. 9(1), 39-46, 1977.

Mourant, R.R. and Rockwell, T.H. Mapping eye movement patterns to the visual scene in driving: an exploratory study. Human Factors. 12(1), 81-87, 1970.

2.15 Haworth-Newman Avionics Display Readability Scale

General description - The Haworth-Newman Avionics Display Readability Scale (see Figure 2) is based on the Cooper-Harper Rating Scale (see Figure 6). As such it has a three-level deep branching that systematically leads to a rating of 1 (excellent) to 10 (major deficiencies).

Strengths and limitations - The scale is easy to use. It has been validated in a limited systematic degradation (i.e., masking) of avionics symbology (Chiappetti, 1994). Recommendations from that validation study include: 1) provide a more precise definition of readability, 2) validate the scale using better trained subjects, 3) use more realistic displays, and 4) use display resolution, symbol luminance, and symbol size to degrade readability.

Data requirements - Subjects must have a copy of the scale in front of them during rating.

Thresholds - 1 (excellent) to 10 (major deficiencies).

Source -

Chiappetti, C.F. Evaluation of the Haworth-Newman Avionics Display Readability Scale. Monterey, CA: Naval Postgraduate School Thesis, September 1994.

2.16 Landing Performance Score

General description - Landing Performance Score (LPS) is a score derived from the multiple regression of the following variables: number of landings per pilot, log book scorings, environmental data (weather, sea state, etc.), aircraft data (type and configuration), carrier data (ship size, visual landing aids, accident rate, etc.), boarding and bolster rate, intervals between landings, mission type and duration, and flying cycle workload estimate. LPS was developed for Navy carrier landings (Brictson, 1977).

Strengths and limitations - LPS distinguished night and day carrier landings (Brictson, 1974).

Data requirements - LPS requires the use of regression techniques.

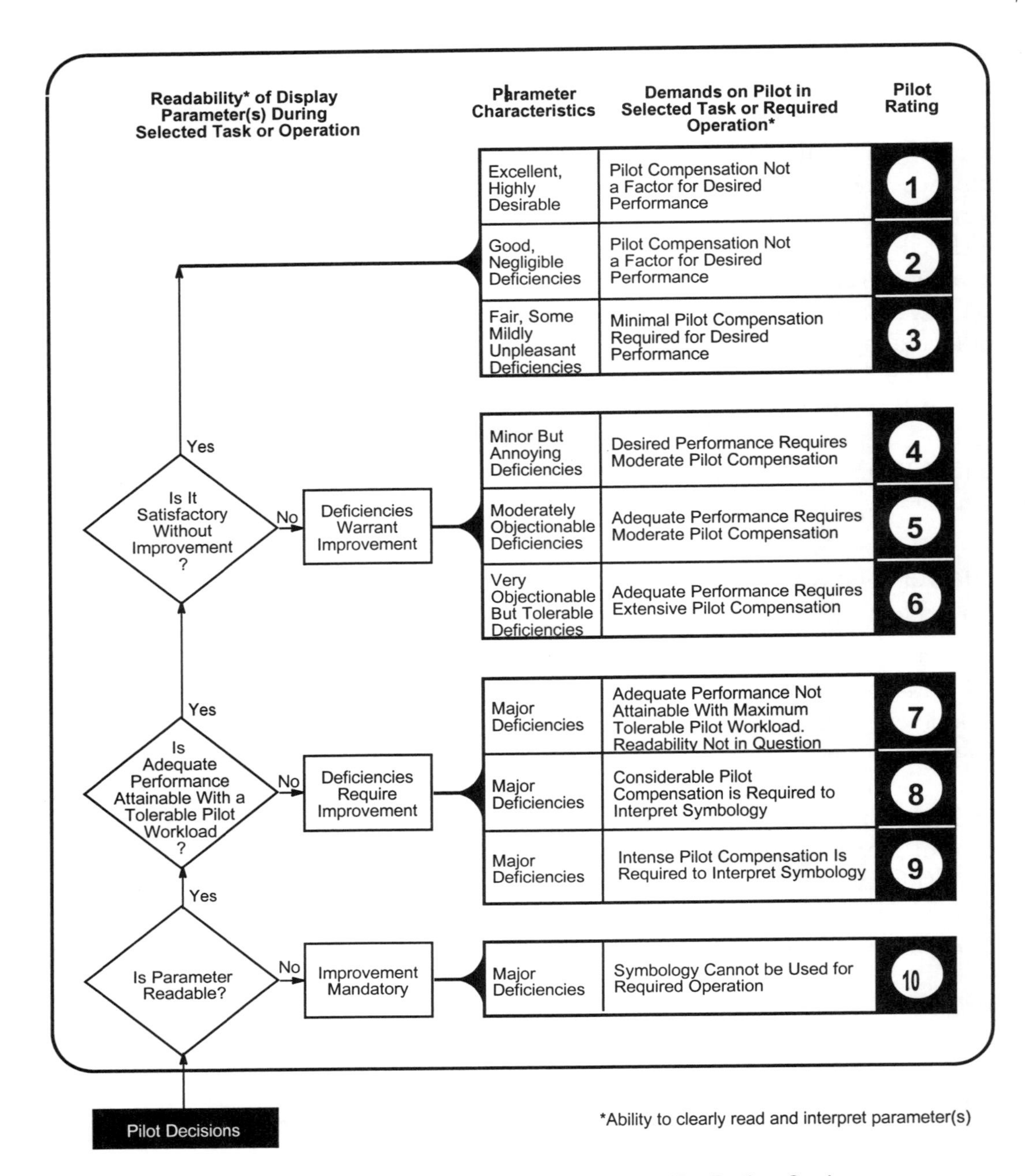

FIG. 2. Haworth-Newman Display Readability Rating Scale (from Haworth, 1993 cited in Chiappetti, 1994)

Thresholds - Not stated.

Sources -

Brictson, C.A. Pilot landing performance under high workload conditions. In A.N. Nickolson (Ed.) Simulation and study of high workload operations (AGARD-CP-146). Neuilly-sur-Seine, France: AGARD; 1974.

Brictson, C.A. Methods to assess pilot workload and other temporal indicators of pilot performances effectiveness. In R. Auffret (Ed.) Advisory Group for Aerospace Research and Development (AGARD) Conference Proceedings Number 217, AGARD-CP-217, B9-7 to B9-10; 1977.

2.17 Lookpoint

General description - Lookpoint is "the current coordinates of where the pilot is looking during any one thirtieth of a second" (Harris, Glover, and Spady, 1986, p. 38). Lookpoint is usually analyzed by either real-time viewing of lookpoint superimposed on the instrument panel or examination of time histories.

Strengths and limitations - Real-time observation of lookpoint efficiently informs the researcher about scanning behavior as well as helps to identify any calibration problems. Analysis of time histories of lookpoint provides information such as average dwell time, dwell percentage, dwell time, fixation, fixations per dwell, one- and two-way transitions, saccades, scans, transition, and transition rate. Such information is useful in (1) arranging instruments for optimum scanning, (2) assessing the time required to assimilate information from each display, (3) estimating the visual workload associated with each display, and task criticality (i.e., blink rate decreases as task criticality increases; Stern, Walrath, and Goldstein, 1984). Corkindale (1974) reported significant differences between working conditions in the percentage of time looking at a HUD. Spady (1977) reported different scanning behavior during approach between manual (73% of time on flight director, 13% on airspeed) and autopilot with manual throttle (50% on flight director, 13% on airspeed).

Limitations include: (1) inclusion of an oculometer into the workplace or simulator, (2) requirement for complex data-analysis software, (3) lack of consistency, (4) difficulty in interpreting the results, and (5) lack of sensitivity of average dwell time. The first limitation is being overcome by the development of miniaturized oculometers, the second by the availability of standardized software packages, the third by collecting enough data to establish a trend, the fourth by development of advanced analysis techniques, and the fifth by use of the dwell histogram.

Data requirements - The oculometer must be calibrated and its data continuously monitored to ensure that it is not out of track. Specialized data reduction and analysis software is required.

Thresholds - Not stated.

Sources -

Corkindale K.G.G. A flight simulator study of missile control performance as a function of concurrent workload. Proceedings of Simulation and Study of High Workload (AGARD-CP-146); 1974.

Harris, R.L., Glover, B.J., and Spady, A.A. Analytical techniques of pilot scanning behavior and their application (NASA Technical Paper 2525). Hampton, VA: NASA Langley; July 1986.

Spady, A.A. Airline pilot scanning behavior during approaches and landing in a Boeing 737 simulator. Proceedings of Guidance and Control Design Considerations for Low Altitude and Terminal Area Flight (AGARD-CP-240); 1977.

Stern, J.A., Walrath, L.C., and Goldstein, R. The indigenous eyeblink. Psychophysiology. 21(1): 22-23; 1984.

2.18 Marking Speed and Errors

General description - Maddox and Turpin (1986) used speed, errors, and error category to evaluate performance using mark-sensed forms. Error categories were: (1) multiple entries in the same row or column, (2) substituting an incorrect number, (3) transposing two or more numbers, and (4) omitting one or more numbers.

Strengths and limitations - None of these measures were affected by number orientation (horizontal or vertical), number ordering (bottom-to-top or top-to-bottom), and handedness of users (Maddox and Turpin, 1986).

Data requirements - Start and stop times must be recorded.

Thresholds - The average marking speed was 4.74 s for a five-digit number; the range of percent errors were 2.2 to 6.3% with an average of 4% per subject (Maddox and Turpin, 1986). Percent of total errors by error category were: (1) 69.3%, (2) 17.3%, (3) 12%, and (4) 1.4% (Maddox and Turpin, 1986).

Source -

Maddox, M.E. and Turpin, J.A. The effect of number ordering and orientation on marking speed and errors for mark-sensed labels. Human Factors. 28(4): 401-405; 1986.

2.19 Mental Arithmetic

General description - Subjects are asked to solve visually or aurally presented mathematical equations without using paper, calculator, or computer.

Strengths and limitations - Ruffell-Smith (1979) compared performance of 20 three-person airline crews on a heavy versus a light two-leg flight. They reported more errors in computation during the heavy workload condition.

Mertens and Collins (1986) reported that performance of this task was not related to age (30 to 39 versus 60 to 69 years old). However, performance degraded as a function of sleep deprivation and altitude (0 versus 3810 m).

Data requirements - Subject responses must be scored immediately to avoid errors in matching problem and answer.

Thresholds - Not stated.

Sources–

Mertens, H.W. and Collins, W.E. The effects of age, sleep deprivation, and altitude on complex performance. Human Factors. 28(5): 541-551; 1986.

Ruffell-Smith, H.P. A simulator study of the interaction of pilot workload with errors, vigilance, and decisions. Moffett Field, CA: Ames Research Center, NASA TM 78432, January 1979.

2.20 Movement Time

General description - Arnaut and Greenstein (1990) provided three definitions of movement time. "Gross movement was defined as the time from the initial touch on the [control] . . . to when the cursor first entered the target. Fine adjustment was the time from the initial target entry to the final lift-off of the finger from the [control]. . . . Total movement time was the sum of these two measures" (p. 655).

Strengths and limitations - Arnaut and Greenstein (1990) reported significant increases in gross movements and significant decreases in fine adjustment times for larger (120 mm) than smaller (40 mm) touch tablets. Total movement times were significantly longer for the largest and smallest touch tablets than for the intermediate size tablets (60,

80, or 100 mm). For a trackball, gross and total movement times were significantly longer for a longer distance (160 or 200 mm) than the shortest distance (40 mm). In a second experiment, there were no significant differences in any of the movement times for the touch tablet with and without a stylus. All three measures were significantly different between different display amplitudes and target widths. "In general performance was better with the smaller display amplitudes and the larger display target widths" (p. 660).

Hancock and Caird (1993) reported that movement time increased as the shrink rate of a target increased. Further movement time decreased as path length increased.

Data requirements - The time at both the beginning and the end of the movement must be recorded.

Thresholds - Movements times below 50 msec are very uncommon.

Sources–

Arnaut, L.Y. and Greenstein, J.S. Is display/control gain a useful metric for optimizing an interface? Human Factors. 32(6), 651-663; 1990.

Hancock, P.A. and Caird, J.K. Experimental evaluation of a model of mental workload. Human Factors. 35(3), 413-419; 1993.

2.21 Nieva, Fleishman, and Rieck's Team Dimensions

General description - Nieva, Fleishman, and Rieck (1985) defined five measures of team performance: (1) matching number resources to task requirements, (2) response coordination, (3) activity pacing, (4) priority assignment among tasks, and (5) load balancing.

Strengths and limitations - The five measures are an excellent first step in developing measures of team performance, but specific metrics must be developed and tested.

Data requirements - The following group characteristics must be considered when using these measures of team performance: (1) group size, (2) group cohesiveness, (3) intra- and inter-group competition and cooperation, (4) communication, (5) standard communication nets, (6) homogeneity/heterogeneity in personality and attitudes, (7) homogeneity/heterogeneity in ability, (8) power distribution within the group, and (9) group training.

Thresholds - Not stated.

Source -

Nieva, V.F., Fleishman, E.A., and Rieck, A. Team dimensions: their identity, their measurement and their relationships. Research Note 85-12. Alexandria, VA: Army Research Institute for the Behavioral and Social Sciences; January 1985.

2.22 Performance Evaluation Tests for Environmental Research (PETER)

General description - The PETER test battery is made up of 26 tests: (1) aiming, (2) arithmetic, (3) associative memory, (4) Atari air combat maneuvering, (5) Atari antiaircraft, (6) choice RT: 1-choice, (7) choice RT: 4-choice, (8) code substitution, (9) flexibility of closure, (10) grammatical reasoning, (11) graphic and phonemic analysis, (12) letter classification: name, (13) letter classification: category, (14) manikin, (15) Minnesota rate of manipulation, (16) pattern comparison, (17) perceptual speed, (18) search for typos in prose, (19) spoke control, (20) Sternberg item recognition: positive set 1, (21) Sternberg item recognition: positive set 4, (22) Stroop, (23) tracking: critical, (24) tracking: dual critical, (25) visual contrast sensitivity, and (26) word fluency (Kennedy, 1985).

Strengths and limitations - Tests were selected for the PETER battery on the following criteria: (1) administration time, (2) total stabilization time, and (3) reliability.

Data requirements - Each PETER test has been programmed for a NEC PC 8201A.

Thresholds - Not stated.

Source -

Kennedy, R.S. A portable battery for objective, non-obtrusive measures of human performance (NASA-CR-171868). Pasadena, CA: Jet Propulsion Laboratory; 1985.

2.23 Pilot Performance Index

General description - Based on subject matter experts, Stein (1984) developed a list of performance variables and associated performance criteria for an air transport mission. The list was subsequently reduced by eliminating those performance measures that did not distinguish experienced from novice pilots (see Table 3). This collection of performance measures was called the Pilot Performance Index (PPI).

TABLE 3.
Pilot Performance Index Variable List

Takeoff	Initial Approach
Pitch Angle	Heading
Climb	Manifold Left
Heading	Manifold Right
Airspeed	Bank Angle
Enroute	Final Approach
Altitude	Heading
Pitch Angle	Gear Position
Heading	Flap Position
Course Deviation Indicator	Course Deviation Indicator
Omni Bearing Sensor	from Stein (1984) p. 20)
Descent	
Heading	
Airspeed	
Bank Angle	
Course Deviation Indicator	
Omni Bearing Sensor	

Strengths and limitations - The PPI provides objective estimates of performance and can distinguish experienced and novice pilots. It does not measure the amount of effort being applied by the pilot, however.

Data requirements - The variables in the PPI are aircraft- and mission-specific.

Thresholds - Not stated.

Source -

Stein, E.S. The measurement of pilot performance: A master-journeyman approach (DOT/FAA/CT-83/15). Atlantic City, NJ: Federal Aviation Administration Technical Center; May 1984.

2.24 Reaction Time

General description - RT is the time elapsed between stimulus onset and response onset. The stimulus is usually a visually presented number requiring a manual key press, but only stimulus and any input or output mode are possible.

Strengths and limitations - RT may measure the duration of mental processing stages (Donders, 1969). RT is a reliable measure (e.g., split-half reliabilities varying between 0.81 and 0.92, AGARD, 9189, p. 12). However, Vidulich (1991) reported a test-retest reliability of +0.39 of a visual choice RT task.

RT is sensitive to physiological state such as fatigue/sleep deprivation, aging, brain damage, and drugs (Boer, Ruzius, Minpen, Bles, and Janssen, 1984; Frowein, 1981; Frowein, Gaillard, and Varey, 1981; Frowein, Reitsma, and Aquarius, 1981; Gaillard, Gruisen, and de Jong, 1986; Gaillard, Rozendaal, and Varey, 1983; Gaillard, Varey, and Ruzius, 1985; Logsdon, Hochhaus, Williams, Rundell, and Maxwell, 1984; Moraal, 1982; Sanders, Wijnen, and van Arkel, 1982; Steyvers, 1987).

Coury, Boulette, and Smith (1989) also reported significant decreases in RT over time. Harris, Hancock, Arthur, and Caird (1995) reported significant decreases in response time as time on task increased. Pigeau, Angus, O'Neill, and Mack (1995) measured response time of air defense operators to detect incoming aircraft. There was a significant interaction of shift and time on task. Subjects working the midnight shift had longer RTs in the 60/60 minute work/rest schedule than those in the evening shift. There was also a shift by zone significant interaction: longer RTs for midnight shift for northern regions (i.e., low air traffic). Finally, between the two sessions, RT increased during the midnight shift but decreased during the evening shift.

Fowler, Elcombe, Kelso, and Portlier (1987) used RT for the first correct response to a five-choice visual RT task to examine the effects of hypoxia. The response time was greater at 82% arterial oxyhaemoglobin saturation than at 84 or 86%. In a related study, Fowler, Mitchell, Bhatia, and Portlier (1989) reported increased RT as a function of inert gas narcosis. However, Boer (1987) suggests that RT performance may require 2000 trials to stabilize. Further, Carter, Krause, and Harbeson (1986) reported that slope was less reliable than RT for a choice RT task.

RT has been used to evaluate auditory, tactile, and visual stimuli.

2.24.1 Auditory Stimuli

Payne and Lang (1991) reported shorter RTs for rapid communication than for conventional visual displays.

RTs were significantly faster for target words in natural than in synthetic speech (Ralston, Pisoni, Lively, Greene, and Mullennix, 1991).

Akamatsu, MacKenzie, and Hasbroucq (1995) reported that there were no differences in response times associated with the type of feedback (normal, auditory, color, tactile, and combined) provided by a computer mouse.

Begault and Pittman (1996) used detection time to compare conventional versus 3-D audio warnings of aircraft traffic. Detection time was significantly shorter (500 msec) for the 3-D audio display.

2.24.2 Tactile Stimuli

In moving-base simulators, RT to crosswind disturbances is significantly shorter when physical-motion cues were present than when they were not resent (Wierwille, Casali, and Repa, 1983).

2.24.3 Visual Stimuli

Simon and Overmeyer (1984) used RT to evaluate the effects of redundant visual cues.

Wierwille, Rahimi, and Casali (1985) reported that RT was significantly affected by the difficulty of a mathematical problem solving task. They defined RT as the time from problem presentation to a correct response.

RT increases as stimulus complexity (1, 2, or 4 vertical lines) increases and interstimulus interval (100, 300, 500, 700, 900, and 1,100 ms) decreases (Aykin, Czaja, and Drury, 1986).

Mertens and Collins (1986) used RT to warning light changes in visual pointer position and successive presentation of targets to evaluate the effects of age (30 to 39 versus 60 to 69 years old), altitude (ground versus 3810 m), and sleep (permitted versus deprived). RT was standardized and then transformed so that better performance was associated with a higher score. Older persons had lower scores in the light task than younger persons. Sleep deprivation decreased performance on all three RT tasks; altitude did not significantly affect performance on either task.

Remington and Williams (1986) used RT to measure the efficiency with which helicopter situation display symbols could be located and identified. RTs for numeric symbols were significantly shorter than for graphic symbols. Negative trials (target not present) average 120 ms longer than positive trials (target present), however, there were more errors on positive than on negative trials.

Downing and Sanders (1987) reported longer response times in simulated control room emergencies with mirror-image panels than non-mirror image panels.

RT has also been useful in discriminating display utility (Nataupsky and Crittenden, 1988).

Koelega, Brinkman, Hendriks, and Verbaten (1989) reported no significant differences in RT between four types of visual vigilance tasks "1) the cognitive Continuous Performance Test (CPT), in which the target was the sequence of the letters AX; 2) a visual version of the cognitive Bakan task in which a target was defined as three successive odd but unequal digits; 3) an analogue of the Bakan using nondigital stimuli; and 4) a pure sensory task in which the critical signal was a change in brightness" (p. 46).

Boehm-Davis, Holt, Koll, Yastrop, and Peters (1989) reported faster RT to database queries if the format of the database was compatible with the information sought.

Coury, Boulette, and Smith (1989) reported significantly faster RTs to a digital display than to a configural or bargraph display. The task was classification.

Imbeau, Wierwille, Wolf, and Chun (1989) used the time from stimulus presentation to a correct answer to evaluate driver reading performance. These authors reported significantly longer RTs for 7-arcmin characters (3.45 to 4.86s) than for 25-arcmin characters (1.35s).

Kline, Ghali, Kline, and Brown (1990) converted visibility distances of road signs to sight times in seconds assuming a constant travel speed. This measure made differences between icon and text signs very evident.

Korteling (1990) compared visual choice RTs among diffuse brain-injury patients, older males (61 to 73 years old), and younger males (21 to 43 years old). The RTs of the patients and older males were significantly longer than the RTs of the younger males. RTs were significantly different as a function of response-stimulus interval (RSI; RSI = 100 msec, RT = 723 msec; 500, 698; 1250, 713, respectively). There was also a significant RSI by stimulus sequence interaction, specifically, "interfering aftereffects of alternating stimuli decreased significantly with increasing RSI" (p. 99).

Jubis (1990) used RT to evaluate display codes. She reported significantly faster RT for redundant color and shape (x = 2.5 s) and color (x = 2.3 s) coding than for partially redundant color (x = 2.8 s) or shape (x = 3.5 s).

Tzelgov, Henik, Dinstein, and Rabany (1990) used RT to compare two types of stereo picture compression. They reported significant task (faster for object decision than depth decision task), depth (faster for smaller depth differences), size (faster for no size difference between compared objects), and presentation effects. There were also numerous interactions.

Buttigieg and Sanderson (1991) reported significant differences in RT between display formats. There were also significant decreases in RT over the three days of the experiment.

Perrott, Sadralodabai, Saberi, and Strybel (1991) reported significant decreases in RT to a visual target when spatially correlated sounds were presented with the visual targets.

Fisk and Jones (1992) reported significant effects of search consistency (varying the ratio of consistent (all words were targets) to inconsistent words: 8:0, 6:2, 4:4, 2:6, and 0:8; shorter RTs with increased consistency) and practice (shorter RT from trial 1 to 12) on correct RT.

McKnight and Shinar (1992) reported significantly shorter brake RT in following vehicles when the forward vehicle had center high-mounted stop lamps. Park and Lee (1992) reported RT in a computer-aided aptitude task did not predict performance of flight trainees.

Yeh and Silverstein (1992) measured RT as subjects made spatial judgments to a simplified aircraft landing display. They reported significantly shorter RT for targets high front versus low back portions of the visual field of the display, for larger altitude separations between targets, for the 45-degree versus the 15-degree viewing orientation, and with the addition of binocular disparity. Krantz, Silverstein, and Yeh (1992) went one step farther and developed a mathematical model to predict RT as a function of spatial frequency, forward field of view/display luminance mismatch, and luminance contrast.

Elvers, Adapathya, Klauer, Kancler, and Dolan (1993) reported significant decreases in RT as a function of practice. This was especially true for a volume estimation versus a distance estimation task.

Hancock and Caird (1993) reported a significant decrease in RT as the length of a path from the cursor to the target increased.

Kimchi, Gopher, Rubin, and Raij (1993) reported significantly shorter RTs for a local-directed rather than a global-directed task in a focused than in a divided attention condition. Significantly shorter RTs were reported for the global-directed than the local-directed task but only in the divided attention condition.

Lee and Fisk (1993) reported faster RTs in a visual search task if the consistency of the stimuli remained 100% than if it did not (67, 50, or 33% consistent).

Briggs and Goldberg (1995) used response time to evaluate armored tank recognition ability. There were significant differences between subjects, presentation time (shorter RTs as presentation time increased), view (flank view faster than frontal view), and model (M1 fastest, British Challenger slowest). There were no significant effects of component vs. friend or foe.

Pigeau, Angus, O'Neill, and Mack (1995) measured response time of air defense operators to detect incoming aircraft. Significantly longer RTs occurred when the same geographical area was displayed in two rather than four zones.

Kerstholt, Passenier, Houltuin, and Schuffel (1996) reported a significant increase in detection time with simultaneous targets. There was also a significant increase in detection time for subsequent targets (first versus second versus third) in two complex target conditions.

Murray and Caldwell (1996) reported significantly longer RTs as the number (1, 2, 3) of displays to be monitored increased and number (1, 2, 3) of display figures increased. There were also longer RTs to process independent rather than redundant images.

2.24.4 Related Measures

Related measures include detection time and recognition time. Detection time is defined as the onset of a target presentation until a correct detection is made. Recognition time is the onset of the target presentation until a due target is correctly recognized (Norman and Ehrlich, 1986).

Bemis, Leeds, and Winer (1988) did not find any significant differences in threat detection times between conventional and perspective radar displays. Response time to select the interceptor nearest to the threat was significantly shorter with the perspective display.

Damos (1985) used a variant of RT, specifically, the average interval between correct responses, (CRI). CRI includes the time to make incorrect responses. CRI was sensitive to variations in stimulus mode and trial.

Data requirements - Start time of the stimulus and response must be recorded to the nearest msec.

Thresholds - RTs below 50 msec are very uncommon.

Sources -

Akamatsu, M., MacKenzie, I.S., and Hasbroucq, T. A comparison of tactile, auditory, and visual feedback in a pointing task using a mouse-type device. Ergonomics. 38(4), 816-827, 1995.

Aykin, N., Czaja, S.J., and Drury, C.G. A simultaneous regression model for double stimulation tasks. Human Factors. 28(6), 633-643; 1986.

Begault, D.R. and Pittman, M.T. Three-dimensional audio versus head-down traffic alert and collision avoidance system displays. International Journal of Aviation Psychology. 6(1), 79-93; 1996.

Bemis, S.V., Leeds, J.L., and Winer, E.A. Operator performance as a function of type of display: conventional versus perspective. Human Factors. 30(2), 163-169; 1988.

Boehm-Davis, D.A., Holt, R.W., Koll, M., Yastrop, G., and Peters, R. Effects of different data base formats on information retrieval. Human Factors. 31(5), 579-592; 1989.

Boer, L.C. Psychological fitness of Leopard I-V crews after a 200-km drive (Report Number 1ZF 1987-30). Soesterberg, Netherlands: TNO Institute for Perception; 1987.

Boer, L.C., Ruzius, M.H.B., Minpen, A.M., Bles, W., and Janssen, W.H. Psychological fitness during a maneuver (Report Number 1ZF 1984-17). Soesterberg, Netherlands: TNO Institute for Perception; 1984.

Briggs, R.W. and Goldberg, J.H. Battlefield recognition of armored vehicles. Human Factors. 37(3): 596-610; 1995.

Buttigieg, M.A. and Sanderson, P.M. Emergent features in visual display design for two types of failure detection tasks. Human Factors. 33(6), 631-651; 1991.

Carter, R.C., Krause, M., and Harbeson, M.M. Beware the reliability of slope scores for individuals. Human Factors. 28(6): 673-683; 1986.

Coury, B.G., Boulette, M.D., and Smith, R.A. Effect of uncertainty and diagnosticity on classification of multidimensional data with integral and separable displays of system status. Human Factors. 31(5), 551-569; 1989.

Damos, D. The effect of asymmetric transfer and speech technology on dual-task performance Human Factors. 27(4), 409-421; 1985.

Donders, F.C. On the speed of mental processes . In W.G. Koster (Ed.) Attention and performance (pp. 412-431). Amsterdam: North Holland; 1969.

Downing, J.V. and Sanders, M.S. The effects of panel arrangement and focus of attention on performance. Human Factors. 29(5), 551-562, 1987.

Elvers, G.C., Adapathya, R.S., Klauer, K.M., Kancler, D.E., and Dolan, N.J. Effects of task probability on integral and separable task performance. Human Factors. 35(4), 629-637, 1993.

Fisk, A.D. and Jones, C.D. Global versus local consistency: effects of degree of within-category consistency on performance and learning. Human Factors. 34(6), 693-705; 1992.

Fowler, B., Elcombe, D.D., Kelso, B., and Portlier, G. The thresholds for hypoxia effects on perceptual-motor performance. Human Factors. 29(1), 61-66; 1987.

Fowler, B., Mitchell, I., Bhatia, M., and Portlier, G. Narcosis has additive rather than interactive effects on discrimination reaction time. Human Factors. 31(5), 571-578; 1989.

Frowein, H.W. Selective drug effects on information processing. Dissertatie, Katholieke Hogeschool, Tilburg; 1981.

Frowein, H.W., Gaillard, A.W.K., and Varey, C.A. EP components, visual processing stages, and the effect of a barbiturate. Biological Psychology. 13, 239-249; 1981.

Frowein, H.W., Reitsma, D., and Aquarius, C. Effects of two counteractivity stresses on the reaction process. In J. Long and A.D. Baddeley (Eds.) Attention and performance. Hillsdale, NJ: Erlbaum; 1981.

Gaillard, A.W.K., Gruisen, A., and de Jong, R. The influence of loratadine (sch 29851) on human performance (Report Number IZF 1986-C-19). Soesterberg, Netherlands: TNO Institute for Perception; 1986.

Gaillard, A.W.K., Rozendaal, A.H., and Varey, C.A. The effects of marginal vitamin-deficiency on mental performance (Report Number IZF 1983-29). Soesterberg, Netherlands: TNO Institute for Perception; 1983.

Gaillard, A.W.K., Varey, C.A., and Ruzius, M.H.B. Marginal vitamin deficiency and mental performance (Report Number 1ZF 1985-22). Soesterberg, Netherlands: TNO Institute for Perception; 1985.

Hancock, P.A. and Caird, J.K. Experimental evaluation of a model of mental workload. Human Factors. 35(3), 413-429; 1993.

Harris, W.C., Hancock, P.A., Arthur, E.J., and Caird, J.K. Performance, workload, and fatigue changes associated with automation. International Journal of Aviation Psychology. 5(2), 169-185; 1995.

Imbeau, D., Wierwille, W.W., Wolf, L.D., and Chun, G.A. Effects of instrument panel luminance and chromaticity on reading performance and preference in simulated driving. Human Factors. 31(2), 147-160; 1989.

Jubis, R.M. Coding effects on performance in a process control task with uniparameter and multiparameter displays. Human Factors. 32(3), 287-297; 1990.

Kerstholt, J.H., Passenier, P.O., Houltuin, K., and Schuffel, H. The effect of a priori probability and complexity on decision making in a supervisory task. Human Factors. 38(1), 65-78; 1996.

Kimchi, R., Gopher, D., Rubin, Y., and Raij, D. Performance under dichoptic versus binocular viewing conditions: effects of attention and task requirements. Human Factors. 35(1), 35-56; 1993.

Kline, T.J.B., Ghali, L.M., Kline, D., and Brown, S. Visibility distance of highway signs among young, middle-aged, and older observers: Icons are better than text. Human Factors. 32(5), 609-619; 1990.

Koelega, H.S., Brinkman, J., Hendriks, L. and Verbaten, M.N. Processing demands, effort, and individual differences in four different vigilance tasks. Human Factors. 31(1), 45-62; 1989.

Korteling, J.E. Perception-response speed and driving capabilities of brain-damaged and older drivers. Human Factors. 32(1), 95-108; 1990.

Krantz, J.H., Silverstein, L.D., and Yeh, Y. Visibility of transmissive liquid crystal displays under dynamic lighting conditions. Human Factors. 34(5), 615-632; 1992.

Lee, M.D. and Fisk, A.D. Disruption and maintenance of skilled visual search as a function of degree of consistency. Human Factors. 35(2), 205-220; 1993.

Logsdon, R., Hochhaus, L., Williams, H.L., Rundell, O.H., and Maxwell, D. Secobarbital and perceptual processing. Acta Psychologica. 55, 179-193; 1984.

McKnight, A.J. and Shinar, D. Brake reaction time to center high-mounted stop lamps on vans and trucks. Human Factors. 34(2), 205-213; 1992.

Mertens, H.W. and Collins, W.E. The effects of age, sleep deprivation, and altitude on complex performance. Human Factors. 28(5), 541-551; 1986.

Moraal, J. Age and information processing: an application of Sternberg's additive factor method (Report Number 1ZF 1982-18). Soesterberg, Netherlands: TNO Institute for Perception; 1982.

Murray, S.A. and Caldwell, B.S. Human performance and control of multiple systems. Human Factors. 38(2), 323-329; 1996.

Nataupsky, M. and Crittenden, L. Stereo 3-D and non-stereo presentations of a computer-generated pictorial primary flight display with pathway augmentation. Proceedings of the 9th AIAA/IEEE Digital Avionics Systems Conference; 1988.

Norman, J. and Ehrlich, S. Visual accommodation and virtual image displays: target detection and recognition. Human Factors. 28(2), 135-151; 1986.

Park, K.S. and Lee, S.W. A computer-aided aptitude test for predicting flight performance of trainees. Human Factors. 34(2); 189-204; 1992.

Payne, D.G. and Lang, V.A. Visual monitoring with spatially versus temporally distributed displays. Human Factors. 33(4), 443-458, 1991.

Perrott, D.R., Sadralodabai, T., Saberi, K., and Strybel, T.Z. Aurally aided visual search in the central visual field: Effects of visual load and visual enhancement of the target. Human Factors. 33(4), 389-400; 1991.

Pigeau, R.A., Angus, R.G., O'Neill, P., and Mack, I. Vigilance latencies to aircraft detection among NORAD surveillance operators. Human Factors. 37(3), 622-634; 1995.

Ralston, J.V., Pisoni, D.B., Lively, S.E., Greene, B.G., and Mullennix, J.W. Comprehension of synthetic speech produced by rule: word monitoring and sentence-by-sentence listening times. Human Factors. 33(4), 471-491; 1991.

Remington, R. and Williams, D. On the selection and evaluation of visual display symbology: factors influencing search and identification times. Human Factors. 28(4), 407-420; 1986.

Sanders, A.F., Wijnen, J.I.C., and van Arkel, A.E. An additive factor analysis of the effects of sleep-loss on reaction processes. Acta Psychologica. 51, 41-59; 1982.

Simon, J.R. and Overmeyer, S.P. The effect of redundant cues on retrieval time. Human Factors. 26(3), 315-321; 1984.

Steyvers, F.J.J.M. The influence of sleep deprivation and knowledge of results on perceptual encoding. Acta Psychologica. 66, 173-178; 1987.

Tzelgov, J., Henik, A., Dinstein, I., and Rabany, J. Performance consequence of two types of stereo picture compression. Human Factors. 32(2); 173-182; 1990.

Wierwille, W.W., Casali, J.G., and Repa, B.S. Driver steering reaction time to abrupt-onset crosswinds, as measured in a moving-base driving simulator. Human Factors. 25(1), 103-116; 1983.

Wierwille, W.W., Rahimi, M., and Casali, J.G. Evaluation of 16 measures of mental workload using a simulated flight task emphasizing mediational activity. Human Factors. 27(5), 489-502; 1985.

Vidulich, M.A. The Bedford Scale: Does it measure spare capacity? Proceedings of the 6th International Symposium on Aviation Psychology. 2, 1136-1141; 1991.

Yeh, Y. and Silverstein, L.D. Spatial judgments with monoscopic and stereoscopic presentation of perspective displays. Human Factors. 34(5), 583-600; 1992.

2.25 Reading Speed

General description - Reading speed is the number of words read divided by the reading time interval. Reading speed is typically measured in words per minute.

Strengths and limitations - Cushman (1986) reported that reading speeds tend to be slower for negative than for positive images. Since there may be a speed accuracy tradeoff, Cushman (1986) also calculated overall reading performance (reading speed x percentage of reading comprehension questions answered correctly).

Gould and Grischkowsky (1986) reported that reading speed decreases as visual angles increase over 24.3 degrees. Gould, Alfaro, Barnes, Finn, Grischkowsky, and Minuto (1987) reported in a series of ten experiments that reading speed is slower from CRT displays than from paper. Gould, Alfaro, Finn, Haupt, and Minuto (1987) concluded on the basis of six studies that reading speed was equivalent on paper and CRT if the CRT displays contained "character fonts that resemble those on paper (rather than dot matrix fonts, for example), that have a polarity of dark characters on a light background, that are anti-aliased (i.e., contain grey level), and that are shown on displays with relatively high resolution (e.g., 1000 x 800)." (p. 497).

Chen, Chan, and Tsoi (1988) used the average reading rate, in words per minute (wpm) to evaluate the effects of window size (20 versus 40 character) and jump length (i.e., number of characters that a message is advanced horizontally) of a visual display. These authors reported that reading rate was significantly less for one-jump (90-91 wpm) than for five- (128 wpm) and nine-jump (139-144) conditions. Reading rate was not significantly affected by window size, however.

Lachman (1989) used the inverse of reading time, i.e., reading rate to evaluate the effect of presenting definitions concurrently with text on a CRT display. There was a significantly higher reading rate for the first 14 screens read than for the second 14 screens.

Jorna and Snyder (1991) reported equivalent reading speeds for hard copy and soft copy displays if the image qualities are similar.

Data requirements - The number of words and the duration of the reading interval must be recorded.

Thresholds - Cushman (1986) reported the following average words/minute: paper = 218; positive image microfiche = 210; negative image microfiche = 199; positive image, negative contrast VDT = 216; and negative image, positive contrast VDT = 209.

Sources–

Chen, H., Chan, K., and Tsoi, K. Reading self-paced moving text on a computer display. Human Factors. 30(3), 285-291; 1988.

Cushman, W.H. Reading from microfiche, a VDT, and the printed page: subjective fatigue and performance. Human Factors. 28(1), 63-73; 1986.

Gould, J.D., Alfaro, L., Barnes, V., Finn, R., Grischkowsky, N., and Minuto, A. Reading is slower from CRT displays than from paper. Attempts to isolate a single-variable explanation. Human Factors. 29(3), 269-299; 1987.

Gould, J.D., Alfaro, L., Finn, R., Haupt, B., and Minuto, A. Reading from CRT displays can be as fast as reading from paper. Human Factors. 29(5), 497-517, 1987.

Gould, J.D. and Grischkowsky, N. Does visual angles of a line of characters affect reading speed. Human Factors. 28(2), 165-173; 1986.

Jorna, G.C. and Snyder, H.L. Image quality determines differences in reading performance and perceived image quality with CRT and hard-copy displays. Human Factors. 33(4), 459-469; 1991.

Lachman, R. Comprehension aids for on-line reading of expository text. Human Factors. 31(1), 1-15; 1989.

2.26 Search Time

General description - Search time is the length of time for a user to retrieve the desired information from a database. Lee and MacGregor (1985) provided the following definition.

$$st = r\,(at + k + c)$$

where

st = search time

r = total number of index pages accessed in retrieving a given item

a = number of alternatives per page

t = time required to read one alternative

k = key-press time

c = computer response time (pp. 158-159)

Matthews (1986) defined search time as the length of time for a subject to locate and indicate the position of a target.

Strengths and limitations - Lee and MacGregor (1985) reported that their search time model was evaluated using a videotex information retrieval system. It was useful in evaluating menu design decisions.

Carter, Krause, and Harbeson (1986) reported that RT was more reliable than slope for this task.

Matthews (1986) reported that search time in the current trial was significantly increased if the visual load of the previous trial was high.

Fisher, Coury, Tengs, and Duffy (1989) used search time to evaluate the effect of highlighting on visual displays. Search time was significantly longer when the probability that the target was highlighted was low (0.25) rather than high (0.75).

Harpster, Freivalds, Shulman, and Leibowitz (1989) reported significantly longer search times using a low resolution/addressability ratio (RAR) than using a high RAR or hard copy.

Matthews, Lovasik, and Mertins (1989) reported significantly longer search times for green on back displays (7.71s) than red on black displays (7.14s).

Lovasik, Matthews, and Kergoat (1989) reported significantly longer search times in the first half hour than in the remaining three and a half hours of a visual search task.

Nagy and Sanchez (1992) used mean log search time to investigate the effects of luminance and chromaticity differences between targets and distractors. "Results showed that mean search time increased linearly with the number of distractors if the luminance difference between target and distractors was small but was roughly constant if the luminance difference was large" (p. 601).

Data requirements - User search time can be applied to any computerized database in which the parameters a, c, k, r, and t can be measured.

Thresholds - Not stated.

Sources–

Carter, R.C., Krause, M., and Harbeson, M.M. Beware the reliability of slope scores for individuals. Human Factors. 28(6): 673-683; 1986.

Fisher, D.L., Coury, B.G., Tengs, T.O., and Duffy, S.A. Minimizing the time to search visual displays: The role of highlighting. Human Factors. 31(2), 167-182; 1989.

Harpster, J.K. Freivalds, A., Shulman, G.L., and Leibowitz, H.W. Visual performance on CRT screens and hard-copy displays. Human Factors. 31(3), 247-257; 1989.

Lee, E. and MacGregor, J. Minimizing user search time in menu retrival systems. Human Factors. 27(2), 157-162; 1985.

Lovasik, J.V., Matthews, M.L., and Kergoat, H. Neural, optical, and search performance in prolonged viewing of chromatic displays. Human Factors. 31(3): 273-289; 1989.

Matthews, M.L. The influence of visual workload history on visual performance. Human Factors. 28(6), 623-632; 1986.

Matthews, M.L., Lovasik, J.V., and Mertins, K. Visual performance and subjective discomfort in prolonged viewing of chromatic displays. Human Factors. 31(3): 259-271; 1989.

Nagy, A.L. and Sanchez, R.R. Chromaticity and luminance as coding dimensions in visual search. Human Factors. 34(5), 601-614; 1992.

2.27 Simulated Work and Fatigue Test Battery

General description - The Simulated Work and Fatigue Test Battery was developed by the National Institute for Occupational Safety and Health to assess the effects of fatigue in the workplace. The simulated work is a data entry task. The fatigue test battery has 11 tasks: grammatical reasoning, digit addition, time estimation, simple auditory RT, choice RT, two-point auditory discrimination, response alternation performance (tapping), hand steadiness, the Stanford sleepiness scale, the NPRU adjective checklist, and oral temperature. Two tasks, grammatical reasoning and simple RT, are also performed in dual-task mode.

Strengths and limitations - The Simulated Work and Fatigue Test Battery is portable, brief, easy to administer, and requires little training of the subject. Rosa and Colligan (1988) used the battery to evaluate the effect of fatigue on performance. All tasks showed significant fatigue effects except: data entry, time production, and two-point auditory discrimination.

Data requirements - The microcomputer provides all task stimuli records all data and scores all tasks.

Thresholds - Not stated.

Source –

Rosa, R.R. and Colligan, M.J. Long workdays versus restdays: Assessing fatigue and alertness with a portable performance battery. Human Factors. 30(3), 305-317; 1988.

2.28 Task Load

General description - Task load is the time required to perform a task divided by the time available to perform the task. Values above 1 indicate excessive taskload.

Strengths and limitations - Task load is sensitive to workload in inflight environments. For example, Geiselhart, Schiffler, and Ivey (1976) used task load to identify differences in workload among four types of refueling missions. Geiselhart, Koeteeuw, and Schiffler (1977) used task load to estimate the workload of KC-135 crews. Using this method, these researchers were able to quantify differences in task load between different types of missions and crew positions. Gunning and Manning (1980) calculated the percentage of time spent on each task for three crewmembers during an aerial refueling. They reported the following inactivity percentages by crew position: pilot, 5 percent; copilot, 45 percent; navigator, 65 percent. Task load was high during takeoff, air refueling, and landing.

Stone, Gulick, and Gabriel (1984), however, identified three problems: "(1) It does not consider cognitive or mental activities. (2) It does not take into account variations associated with ability and experience or dynamic, adaptive behavior. (3) It cannot deal with simultaneous or continuous-tracking tasks" (p. 14).

Data requirements - Use of the task load method requires: (1) clear visual and auditory records of pilots in flight and (2) objective measurement criteria for identifying the starts and ends of tasks.

Thresholds - Not stated.

Sources -

Geiselhart, R., Koeteeuw, R.I., and Schiffler, R.J. A study of task loading using a four-man crew on a KC-135 aircraft (Giant Boom) (ASD-TR-76-33). Wright-Patterson Air Force Base, OH: Aeronautical Systems Division; April 1977.

Geiselhart, R., Schiffler, R.J., and Ivey, L.J. A study of task loading using a three man crew on a KC-135 aircraft (ASD-TR-76-19). Wright-Patterson Air Force Base, OH: Aeronautical Systems Division; October 1976.

Gunning, D. and Manning, M. The measurement of aircrew task loading during operational flights. Proceedings of the Human Factors Society 24th Annual Meeting (pp. 249-252). Santa Monica, CA: Human Factors Society; 1980.

Stone, G., Gulick, R.K., and Gabriel, R.F. Use of task/timeline analysis to assess crew workload (Douglas Paper 7592). Longbeach, CA: Douglas Aircraft Company; 1984.

2.29 Time to Complete

General description - Time to complete is the duration from the subject's first input to the last response (Casali, Williges, and Dryden, 1990).

Strengths and limitations - Time to complete provides a measure of task difficulty but may be traded off for accuracy. Casali, Williges, and Dryden (1990) reported significant effects of speech recognition system accuracy and available vocabulary but not a significant age effect. There were also several interactions.

Frankish and Noyes (1990) used rate of data entry to evaluate data feedback techniques. They reported significantly higher rates for visual presentation and for visual feedback than for spoken feedback.

Adelman, Cohen, Bresnick, Chinnis, and Laskey (1993) used time to complete an aircraft identification task to evaluate expert system interfaces and capabilities. They reported that operators took longer to examine aircraft with the screening rather than with the override interface.

Brand and Judd (1993) reported significant differences in editing time as a function of the angle of hard copy (316.75s for 30 degree, 325.03s for 0 degree; and 371.92s for 90 degree).

Massimino and Sheridan (1994) reported no difference in task competition times between direct and video viewing during teleoperation.

In early work, Burger, Knowles, and Wulfeck (1970) used experts to estimate the time it would take to perform tasks. The estimates were then tested against the real times. Although the correlation between the estimates and the real times was high (+0.98), estimates of the minimum performance times were higher than the actual times and varied widely between judges. The tasks were throw toggle switch, turn rotary switch to a specified value, push toggle, observe and record data, and adjust dial.

Troutwine and O'Neal (1981) reported time was judged shorter on an interesting rather than a boring task but only for subjects with volition. For subjects without volition there was no significant difference in time estimation between the two types of tasks.

Data requirements - The start and end of the task must be well defined.

Thresholds - Minimum time is 30 msec.

Sources -

Adelman, L., Cohen, M.S., Bresnick, T.A., Chinnis, J.O., and Laskey, K.B. Real-time expert system interfaces, cognitive processes, and task performance: An empirical assessment. Human Factors 35(2), 243-261; 1993.

Brand, J.L and Judd, K. Angle of hard copy and text-editing performance. Human Factors. 35(1), 57-70; 1993.

Burger, W.J., Knowles, W.B., and Wulfeck, J.W. Validity of expert judgments of performance time. Human Factors, 12(5); 503-510; 1970.

Casali, S.P., Williges, B.H., and Dryden, R.D. Effects of recognition accuracy and vocabulary size of a speech recognition system on task performance and user acceptance. Human Factors. 32(2); 183-196; 1990.

Frankish, C. and Noyes, J. Sources of human error in data entry tasks using speech input. Human Factors. 32(6); 697-716; 1990.

Massimino, J.J. and Sheridan, T.B. Teleoperator performance with varying force and visual feedback. Human Factors. 36(1), 145-157; 1994.

Troutwine, R. and O'Neal, E.C. Volition, performance of a boring task and time estimation. Perceptual and Motor Skills. 52, 865-866; 1981.

2.30 Time-to-Line-Crossing (TLC)

General description - TLC was developed to enhance preview-predictor models of human driving performance. TLC equals the time for the vehicle to reach either edge of the driving lane. It is calculated from lateral lane position, the heading angle, vehicle speed, and commanded steering angle (Godthelp, Milgram, and Blaauw, 1984).

Strengths and limitations - Godthelp, Milgram, and Blaauw (1984) evaluated TLC in an instrumented car driven on an unused straight, four-lane highway by six male drivers at six different speeds (20, 40, 60, 80, 100, and 120 km/hr) with and without a visor. Based on the results, the authors argued that TLC was a good measure of open-loop driving performance. Godthelp (1986) reported, based on field study data, that TLC described anticipatory steering action during curve driving.

Finnegan and Green (1990) reviewed five studies in which time to change lanes was measured. For these studies, the authors concluded that 6.6 seconds are required for the visual search with a single lane change and 1.5 seconds to complete the lane change.

Data requirements - Lateral lane position, heading angle, vehicle speed, and commanded steering angle must be recorded.

Thresholds - Not stated.

Sources -

Finnegan, P. and Green, P. The time to change lanes: a literature review (UMTRI-90-34). Ann Arbor, MI: The University of Michigan Transportation Research Institute; September 1990.

Godthelp, H. Vehicle control during curve driving. Human Factors. 28(2), 211-221; 1986.

Godthelp, H., Milgram, P., and Blaauw, G.J. The development of a time-related measure to describe driving strategy. Human Factors. 26(3), 257-268; 1984.

2.31 Unified Tri-services Cognitive Performance Assessment Battery (UTCPAB)

General description - The UTCPAB is made up of 25 tests: (1) linguistic processing, (2) grammatical reasoning (traditional), (3) grammatical reasoning (symbolic), (4) two-column addition, (5) mathematical processing, (6) continuous recognition, (7) four-choice serial RT, (8) alpha-numeric visual vigilance, (9) memory search, (10) spatial processing, (11) matrix rotation, (12) manikin, (13) pattern comparison (simultaneous), (14) pattern comparison (successive), (15) visual scanning, (16) code substitution, (17) visual probability monitoring, (18) time wall, (19) interval production, (20) Stroop, (21) dichotic listening, (22) unstable tracking, (23) Sternberg-tracking combination, (24) matching to sample, and (25) item order.

Strengths and limitations - Tests were selected for the UTCPAB based on the following criteria: (1) used in at least one Department of Defense laboratory, (2) proven validity, (3) relevance to military performance, and (4) sensitivity to hostile environments and sustained operations (Perez, Masline, Ramsey, and Urban, 1987).

Data requirements - Each UTCPAB test has been programmed for computer administration. Standardized instructions are used as well as a standardized data file format.

Thresholds - Not stated.

Sources -

Perez, W.A., Masline, P.J., Ramsey, E.G., and Urban, K.E. Unified tri-services cognitive performance assessment battery: review and methodology (AAMRL-TR-87-007). Wright-Patterson Air Force Base, OH: Armstrong Aerospace Medical Research Laboratory; March 1987.

3 Human Workload

Workload has been defined as a set of task demands, as effort, and as activity or accomplishment (Gartner and Murphy 1979). It has been measured as performance (see section 3.1) or as subjective estimates (see section 3.2).

Guidelines for selecting the appropriate workload measure are given in Wierwille, Williges, and Schiflett (1979) and O'Donnell and Eggemeier (1986); for mental workload in Moray (1982). Wierwille and Eggemeier (1993) listed four aspects of measures that were critical: diagnosticity, global sensitivity, transferability, and implementation requirements.

Sources -

Gartner, W.B. and Murphy, M.R. Concepts of workload. In B.O. Hartman and R.E. McKenzie (Eds.) Survey of methods to assess workload. AGARD-AG-246; 1979.

Moray, N. Subjective mental workload. Human Factors. 24(1), 25-40; 1982.

O'Donnell, R.D. and Eggemeier, F.T. Workload assessment methodology. In K.R. Boff, L. Kaufman, and J.P. Thomas (Eds.) Handbook of perception and human performance. New York, NY: John Wiley and Sons; 1986.

Wierwille, W.W., and Eggemeier, F.T. Recommendations for mental workload measurement in a test and evaluation environment. Human Factors. 35(2), 263-281; 1993.

Wierwille, W.W., Williges, R.C., and Schiflett, S.G. Aircrew workload assessment techniques. In B.O. Hartman and R.E. McKenzie (Eds.) Survey of methods to assess workload. AGARD-AG-246; 1979.

3.1 Performance Measures of Workload

Performance of both primary and secondary tasks has been used to measure workload. These measures assume that, as workload increases, the additional processing requirements will degrade performance. O'Donnell and Eggemeier (1986) identified four problems associated with using performance as a measure of workload: (1) underload may enhance performance, (2) overload may result in a floor effect, (3) confounding effects of information-processing strategy, training, or experience, and (4) measures are task specific and cannot be generalized to other tasks.

Meshkati, Hancock, and Rahimi (1990) stated that multiple primary task measures are required when the task is complex or multidimensional. In addition, task measures may be intrusive and may be influenced by factors other than workload, for example, motivation and learning.

Sources -

Meshkati, N., Hancock, P.A., and Rahimi, M. Techniques in mental workload assessment. In J.R. Wilson and E.N. Corlett (Eds.) Evaluation of human work. A practical ergonomics methodology. New York: Taylor and Francis; 1990.

O'Donnell, R.D. and Eggemeier, F.T. Workload assessment methodology. In K.R. Boff, L. Kaufman, and J.P. Thomas (Eds.) Handbook of perception and human performance. New York: Wiley and Sons; 1986.

3.1.1 Aircrew Workload Assessment System

General description - The Aircrew Workload Assessment System (AWAS) is timeline analysis software developed by British Aerospace to predict workload. AWAS requires three inputs: 1) second-by-second description of pilot tasks during flight, 2) demands on each of Wicken's multiple resource theory processing channels, and 3) effects of simultaneous demand on a single channel (Davies, Tomoszek, Hicks, and White (1995).

Strengths and limitations - Davies, et al. (1995) reported a correlation of +0.904 between AWAS workload prediction and errors in a secondary auditory discrimination task. The subjects were two experienced pilots flying a Sea Harrier Simulator.

Data requirements - Second-by-second timeline of pilot tasks, demands on "each information processing channel, and effect of simultaneous demand.

Thresholds - Not stated.

Source -

Davies, A.K., Tomoszek, A., Hicks, M.R., and White, J. AWAS (Aircrew Workload Assessment System): issues of theory, implementation, and validation. In R. Fuller, N. Johnston, and N. McDonald (Eds.) Human factors in aviation operations. Proceedings of the 21st conference of the European Association for Aviation Psychology (EAAP) Volume 3, Chapter 48, 1995.

3.1.2 Control Movements/Unit Time

General description - Control movements/unit time is the number of control inputs made summed over each control used by one operator divided by the unit of time over which the measurements were made.

Strengths and limitations - Wierwille and Connor (1983) stated that this measure was completely sensitive to workload. Their specific measure was the average count per second of inputs into the flight controls (ailerons, elevator, and rudder) in a moving-base, flight simulator. The workload manipulation was pitch stability, wind-gust disturbance, and crosswind direction and velocity.

Zeitlin (1995) developed a driver workload index based on brake actuations per minute plus the log of vehicle speed. This index was sensitive to differences in roadway (rural, city, expressway).

Porterfield (1997), using a similar approach, evaluated the use of the duration of time that an en route air traffic controller was engaged in ground-to-air communications as a measure of workload. He reported a significant correlation (+0.88) between the duration and the Air Traffic Workload Input Technique (ATWIT), a workload rating based on Pilot Objective/Subjective Workload Assessment Technique (POSWAT).

Data requirements - Control movements must be well defined.

Thresholds - Not stated.

Sources -

Porterfield, D.H. Evaluating controller communication time as a measure of workload. The International Journal of Aviation Psychology. 7(2), 171-182, 1997.

Wierwille, W.W. and Connor, S.A. Evaluation of 20 workload measures using a psychomotor task in a moving-base aircraft simulator. Human Factors. 25(1), 1-16, 1983.

Zeitlin, L.R. Estimates of driver mental workload: a long term field trial of two subsidiary tasks. Human Factors. 37(3), 611-621; 1995.

3.1.3 Glance Duration and Frequency

General description - The duration and frequency of glances to visual displays have been used as measures of visual workload. The longer the durations and/or the greater the frequency of glances, the higher the visual workload.

Strengths and limitations - Fairclough, Ashby, and Parkes (1993) used glance duration to calculate the percentage of time that drivers looked at navigation information (a paper map versus an LCD text display), roadway ahead, rear view mirror, dashboard, left-wing mirror, right-wing mirror, left window, and right window. Data were collected in an instrumented vehicle driven on British roads. The authors concluded that this "measure proved sensitive enough to (a) differentiate between the paper map and the LCD/text display and (b) detect associated changes with regard to other areas of the visual scene" (p. 248). These authors warned, however, that reduction in glance durations might reflect the drivers' strategy to cope with the amount and legibility of the paper map.

These authors also used glance duration and frequency to compare two in-vehicle route guidance systems. The data were collected from 23 subjects driving an instrumented vehicle in Germany. The data indicate "as glance frequency to the navigation display increases, the number of glances to the dashboard, rear-view mirror and the left-wing mirror all show a significant decrease" (p. 251). Based on these results, the authors concluded that "glance duration appears to be more sensitive to the difficulty of information update. glance frequency represents the amount of. "visual checking behavior" (p. 251).

Wierwille (1993) concluded from a review of driver visual behavior that such behavior is "relatively consistent" (p. 278).

Data requirements - Record subject's eye position.
Threshold - 0 to infinity.
Sources -

Fairclough, S.H., Ashby, M.C., and Parkes, A.M. In-vehicle displays, visual workload and visibility evaluation. In A.G. Gale, I.D. Brown, C.M. Haslegrave, H.W. Kruysse, and S.P. Taylor (Eds). Vision in Vehicles - IV. Amsterdam: North-Holland; 1993.
Wierwille, W.W. An initial model of visual sampling of in-car displays and controls. In A.G. Gale, I.D. Brown, C.M. Haslegrave, H.W. Kruysse, and S.P. Taylor (Eds.) Vision in Vehicles - IV. Amsterdam: North-Holland; 1993.

3.1.4 Load Stress

General description - Load stress is the stress produced by increasing the number of signal sources that must be attended during a task (Chiles and Alluisi, 1979).
Strengths and limitations - Load stress affects the number of errors made in task performance. Increasing load stress to measure operator workload may be difficult in nonlaboratory settings.
Data requirements - The signal sources must be unambiguously defined.
Thresholds - Not stated.
Source -

Chiles, W.D. and Alluisi, E.A. On the specification of operator or occupational workload with performance-measurement methods. Human Factors. 21(5), 515-528; 1979.

3.1.5 Observed Workload Area

General description - Laudeman and Palmer (1995) developed Observed Workload Area to measure workload in aircraft cockpits. The measure is not based on theory but rather on a logical connection between workload and task constraints. In their words:
"An objectively defined window of opportunity exists for each task in the cockpit. The observed workload of a task in the cockpit can be operationalized as a rating of maximum task importance supplied by a domain expert. Task importance increases during the task window of opportunity as a linear function of task importance versus time. When task windows of opportunity overlap, resulting in an overlap of task functions, the task functions can be combined in an additive manner to produce a composite function that includes the observed workload effects of two or more task functions. We called these composites of two or more task functions observed-workload functions. The dependent measure that we proposed to extract from the observed-workload function was the area under it that we called observed-workload area" (pp. 188-190).
Strengths and limitations - Laudeman and Palmer (1995) reported a significant correlation between the first officers' workload-management ratings and the observed-workload area. This correlation was based on 18 two-person aircrews flying a high-fidelity aircraft simulator. Small observed-workload area was associated with high-workload management ratings. Higher error rates crews had higher observed-workload areas. The technique requires an expert to provide task importance ratings. It also requires well-defined beginnings and ends to tasks.
Thresholds - Not stated.

Source -

Laudeman, I.V. and Palmer, E.A. Quantitative measurement of observed workload in the analysis of aircrew performance. International Journal of Aviation Psychology. 5(2), 187-197; 1995.

3.1.6 Rate of Gain of Information

General description - This measure is based on Hick's Law, which states that RT is a linear function of the amount of information transmitted, (H_t): RT = a + B (H_t) (Chiles and Alluisi, 1979).

Strengths and limitations - Hick's Law has been verified in a complete range of conditions. However, it is limited to only discrete tasks and, unless the task is part of the normal procedures, may be intrusive, especially in nonlaboratory settings.

Data requirements - Rate of gain of information is estimated from RT. Time is typically collected with either mechanical stop watches or software clocks. The first type of clock requires frequent (for example, prior to every trial) calibration; software clocks require a stable and constant source of power.

Thresholds - Not stated.

Source -

Chiles, W.D. and Alluisi, E.A. On the specification of operator or occupational workload with performance-measurement methods. Human Factors. 21(5), 515-528; 1979.

3.1.7 Relative Condition Efficiency

General description - Paas and van Merrienboer (1993) combined ratings of workload with task performance measures to calculate relative condition efficiency. Ratings varied from 1 (very, very low mental effort) to 9 (very, very high mental effort). Performance was measured as percent correct answers to test questions. Relative condition efficiency was calculated "as the perpendicular distance to the line that is assumed to represent an efficiency of zero" (p. 737).

Strengths and limitations - Efficiency scores were significantly different between work conditions.

Data requirements – Not stated.

Thresholds – Not stated.

Source -

Paas, F.G.W.C. and van Merrienboer, J.J.G. The efficency of instructional conditions: An approach to combine mental effort and performance measures. Human Factors. 35(4), 737-743; 1993.

3.1.8 Speed Stress

General description - Speed stress is stress produced by increasing the rate of signal presentation from one or more signal sources.

Strengths and limitations - Speed stress affects the number of errors made as well as the time to complete tasks (Conrad, 1956; Knowles, Garvey, and Newlin, 1953). It may be difficult to impose speed stress on nonlaboratory tasks.

Data requirements - The task must include discrete signals whose presentation rate can be manipulated.

Thresholds - Not stated.

Sources -

Conrad, R. The timing of signals in skill. Journal of Experimental Psychology. 51, 365-370; 1956.

Knowles, W.B., Garvey, W.D., and Newlin, E.P. The effect of speed and load on display-control relationships. Journal of Experimental Psychology. 46, 65-75; 1953.

3.1.9 Secondary Tasks

One of the techniques used most widely to measure workload is the secondary task. This technique requires an operator to perform the primary task within that task's specified requirements and to use any spare attention or capacity to perform a secondary task. The decrement in performance of the secondary task is operationally defined as a measure of workload.

The secondary-task technique has several advantages. First, it may provide a sensitive measure of operator capacity and may distinguish among alternative equipment configurations that are indistinguishable by single-task performance (Slocum, Williges, and Roscoe, 1971). Second, it may provide a sensitive index of task impairment due to stress. Third, it may provide a common metric for comparisons of different tasks.

The secondary-task technique may have one major disadvantage: intrusion on the performance of the primary task (Williges and Wierwille, 1979) and as Rolfe (1971) stated: "The final word, however, must be that the secondary task is no substitute for competent and comprehensive measurement of primary task performance" (p. 146). Vidulich (1989a), however, concluded from two experiments that secondary tasks that do not intrude on primary-task performance are insensitive to primary-task difficulty. Vidulich (1989b) argued that added task sensitivity is directly linked to intrusiveness.

Further, subjects may use different strategies when performing secondary tasks. For example, Schneider and Detweiler (1988) identified seven compensatory activities that are associated with dual-task performance: "1) shedding and delaying tasks and preloading buffers, 2) letting go of high-workload strategies, 3) utilizing noncompeting resources, 4) multiplexing over time, 5) shortening transmissions, 6) converting interference from concurrent transmissions, and 7) chunking of transmissions" (p. 539). In addition, Meshkati, Hancock, and Rahimi (1990) recommend not using secondary task and subjective measures in the same experiment since operators may include secondary-task performance as part of their subjective workload rating.

Damos (1993) analyzed the results of 14 studies in which single- and dual-task performances were evaluated. She concluded that: "the effect sizes associated with both single- and multiple-task measures were both statistically different from 0.0, with the effect size for the multiple-task increases statistically greater than that of the corresponding single task measures. However, the corresponding predictive validities were low." (p. 615).

However, Colle, Amell, Ewry, and Jenkins (1988) developed the method of double trade-off curves to equate performance levels on different secondary tasks. In this method, "two different secondary tasks are each paired with the same primary tasks. A trade-off curve is obtained for each secondary task paired with the primary task" (p. 646).

Knowles (1963) developed a comprehensive set of criteria for selecting a secondary task: (1) noninterference with the primary task, (2) ease of learning, (3) self-pacing,

(4) continuous scoring, (5) compatibility with the primary task, (6) sensitivity, and (7) representativeness.

Sources -

Colle, H., Amell, J.R., Ewry, M.E., and Jenkins, M.L. Capacity equivalence curves: a double trade-off curve method for equating task performance. Human Factors. 30(5), 645-656; 1988.

Damos, D. Using meta-analysis to compare the predictive validity of single- and multiple-task measures of flight performance. Human Factors. 35(4), 615-628; 1993.

Knowles, W.B. Operator loading tasks. Human Factors. 5, 151-161; 1963.

Meshkati, N., Hancock, P.A., and Rahimi, M. Techniques in mental workload assessment. In J.R. Wilson and E.N. Corlett (Eds.) Evaluation of a human work. A practical ergonomics methodology. New York: Taylor and Francis; 1990.

Rolfe, J.M. The secondary task as a measure of mental load. In W.T. Singleton, J.G. Fox, and D. Whitfield (Eds.) Measurement of man at work. London: Taylor and Francis Ltd; 1971.

Slocum, G.K., Williges, B.H., and Roscoe, S.N. Meaningful shape coding for aircraft switch knobs. Aviation Research Monographs. 1 (3), 27-40; 1971.

Schneider, W. and Detweiler, M. The role of practice in dual-task performance: toward workload modeling in a connectionist/control architecture. Human Factors. 30(5), 539-566; 1988.

Vidulich, M.A. Objective measures of workload: Should a secondary task be secondary? Proceedings of the Fifth International Symposium on Aviation Psychology. 802-807, 1989a.

Vidulich, M.A. Performance-based workload assessment: allocation strategy and added task sensitivity. Proceedings of the Third Annual Workshop on Space Operations, Automation, and Robotics (SOAR '89). 329-335; 1989b.

Williges, R.C. and Wierwille, W.W. Behavioral measures of aircrew mental workload. Human Factors. 21, 549-574; 1979.

3.1.9.1 Card Sorting Secondary Task

General description - "The subject must sort playing cards by number, color, and/or suite." (Lysaght, et al., 1989, p. 234).

Strengths and limitations - "Depending upon the requirements of the card sorting rule, the task can impose demands on perceptual and cognitive processes" (Lysaght, et al., 1989, p. 234). Lysaght, et al. (1989) state that dual-task pairing of a primary memory task with a secondary card-sorting task resulted in a decrement in performance in both tasks. Their statement is based on two experiments by Murdock (1965).

Data requirements - The experimenter must be able to record the number of cards sorted and the number of incorrect responses.

Thresholds - Not stated.

Sources -

Lysaght, R.J., Hill, S.G., Dick, A.O., Plamondon, B.D., Linton, P.M., Wierwille, W.W., Zaklad, A.L., Bittner, A.C., and Wherry, R.J. Operator Workload: comprehensive review and evaluation of operator workload methodologies (Technical Report 851). Alexandria, VA: Army Research Institute for the Behavioral and Social Sciences; June 1989.

Murdock, B.B. Effects of a subsidiary task on short-term memory. British Journal of Psychology. 56, 413-419; 1965.

3.1.9.2 Choice RT Secondary Task

General description - "The subject is presented with more than one stimulus and must generate a different response for each one" (Lysaght, et al., 1989, p. 232).

Strengths and limitations - "Visual or auditory stimuli may be employed and the response mode is usually manual. It is theorized that choice RT imposes both central processing and response selection demands" (Lysaght, et al., 1989, p. 232).

Based on nineteen studies that included a choice RT secondary task, Lysaght, et al. (1989) stated that, in dual-task pairings: performance of choice RT, problem solving, and flight simulation primary tasks remained stable; performance of tracking, choice RT, memory, monitoring, driving, and lexical decision primary tasks degraded; and tracking performance improved. Performance of the secondary task remained stable with tracking and driving primary tasks; and degraded with tracking, choice RT, memory, monitoring, problem solving, flight simulation, driving, and lexical decision primary tasks (see Table 4).

Hicks and Wierwille (1979) compared five measures of workload. They manipulated workload by increasing wind gust in a driving simulator. They reported that a secondary RT task was not as sensitive to wind gust as were steering reversals, yaw deviation, subjective opinion rating scales, and lateral deviations. Gawron (1982) reported longer RTs and lower percent correct scores when a four-choice RT task was performed simultaneously then sequentially.

Klapp, Kelly, and Netick (1987) asked subjects to perform a visual, zero-order, pursuit tracking task with the right hand while performing a two-choice, auditory reaction task with the left hand. In the dual-task condition the tracking task was associated with hesitations lasting 333 ms or longer. Degradations in the tracking task were associated with enhancements of the RT task.

Data requirements - The experimenter must be able to record and calculate: mean RT for correct responses, mean (median) RT for incorrect responses, number of correct responses, and number of incorrect responses.

Thresholds - Not stated.

Sources -

Allen, R.W., Jex, H.R., McRuer, D.T., and DiMarco, R.J. Alcohol effects on driving behavior and performance in a car simulator. IEEE Transactions on Systems Man and Cybernetics. SMC-5, 485-505; 1976.

Becker, C.A. Allocation of attention during visual word recognition. Journal of Experimental Psychology: Human Perception and Performance. 2, 556-566; 1976.

Benson, A.J., Huddleston, J.H.F., and Rolfe, J.M. A psychophysiological study of compensatory tracking on a digital display. Human Factors. 7, 457-472; 1965.

Bortolussi, M.R., Hart, S.G., and Shively, R.J. Measuring moment-to-moment pilot workload using synchronous presentations of secondary tasks in a motion-base trainer. Proceedings of the Fourth Symposium on Aviation Psychology. Columbus, OH: Ohio State University; 1987.

Bortolussi, M.R., Kantowitz, B.H., and Hart, S.G. Measuring pilot workload in a motion base trainer. Applied Ergonomics. 17, 278-283; 1986.

TABLE 4.
References Listed by the Effect on Performance of Primary Tasks Paired with a Secondary Choice RT Task

Type	PRIMARY TASK			SECONDARY TASK		
	Stable	Degraded	Enhanced	Stable	Degraded	Enhanced
Choice RT	Becker (1976) Ellis (1973)	Detweile and Lundy (1995)* Gawron (1982)* Schvaneveldt (1969)		Hicks and Wierwille (1979)	Becker (1976) Detweiler and Lundy (1995)* Ellis (1973) Gawron (1982)* Schvaneveldt (1969)	
Driving		Allen, Jex, McRuer, and DiMarco (1976) Brown, Tickner, and Simmonds (1969)		Allen, *et al.* (1976) Drory (1985)*	Brown, *et al.* (1969)	
Flight Simulation	Bortolussi, Hart, and Shively (1987) Bortolussi, Kantowitz, and Hart (1986) Kantowitz, Hart and Bortolussi (1983)* Kantowitz, Hart, Bortolussi, Shively, and Kantowitz (1984)*				Bortolussi, *et al.* (1987) Bortolussi, *et al.* (1986)	
Lexical Decision		Becke (1976)			Becker (1976)	
Memory		Logan (1970)			Logan (1970)	
Monitoring		Smith (1969)			Krol (1971) Smith (1969)	
Problem Solving	Fisher1975a) Fisher (1975b)				Fisher (1975a) Fisher (1975b)	
Tracking		Benson, Huddleston, and Rolfe (1965) Giroud, Laurencelle, and Proteau (1984) Israel, Chesney, Wickens, and Donchin (1980) Israel, Wickens, Chesney, and Donchin (1980) Klapp, Kelly, Battiste, and Dunbar (1984) Wempe and Baty (1968) Klapp, Kelly, and Netick (1987)*			Benson, *et al.* (1965) Damos (1978) Giroud, *et al.* (1984) Israel, *et al.* (1980) Klapp, *et al.* (1984)	Klapp, Kelly, and Netick (1987)*

from Lysaght, et al. (1989) p. 246
**Not included in Lysaght, et al. (1989)*

Brown, I.D., Tickner, A.H., and Simmonds, D.C.V. Interference between concurrent tasks of driving and telephoning. Journal of Applied Psychology. 53, 419-424; 1969.

Damos, D. Residual attention as a predictor of pilot performance. Human Factors. 20, 435-440; 1978.

Detweiler, M. and Lundy, D.H. Effects of single- and dual-task practice an acquiring dual-task skill. Human Factors. 37(1), 193-211; 1995.

Drory, A. Effects of rest and secondary task on simulated truck-driving performance. Human Factors. 27(2), 201-207, 1985.

Ellis, J.E. Analysis of temporal and attentional aspects of movement control. Journal of Experimental Psychology. 99, 10-21; 1973.

Fisher, S. The microstructure of dual task interaction. 1. The patterning of main-task responses within secondary-task intervals. Perception. 4, 267-290; 1975a.

Fisher, S. The microstructure of dual task interaction. 2. The effect of task instructions on attentional allocation and a model of attentional-switching. Perception. 4, 459-474; 1975b.

Gawron, V.J. Performance effects of noise intensity, psychological set, and task type and complexity. Human Factors. 24(2), 225-243; 1982.

Giroud, Y., Laurencelle, L., and Proteau, L. On the nature of the probe reaction-time task to uncover the attentional demands of movement. Journal of Motor Behavior. 16, 442-459; 1984.

Hicks, T.G. and Wierwille, W.W. Comparison of five mental workload assessment procedures in a moving-base during simulator. Human Factors. 21, 129-143, 1979.

Israel, J.B., Chesney, G.L., Wickens, C.D., and Donchin, E. P300 and tracking difficulty: Evidence for multiple resources in dual-task performance. Psychophysiology. 17, 259-273; 1980.

Israel, J.B., Wickens, C.D., Chesney, G.L., and Donchin, E. The event related brain potential as an index of display-monitoring workload. Human Factors. 22, 211-224; 1980.

Kantowitz, B.H., Hart, S.G., and Bortolussi, M.R. Measuring pilot workload in a moving-base simulator I. Asynchronous secondary choice-reaction time task. Proceedings of the Twenty-Seventh Annual Meeting of the Human Factors Society. Santa Monica, CA: Human Factors Society; 1983.

Kantowitz, B.H., Hart, S.G., Bortolussi, M.R., Shively, R.J., and Kantowitz, S.C. Measuring pilot workload in a moving-base simulator II. Building levels of workload. NASA 20th Annual Conference on Manual Control. 2, 373-396; 1984.

Klapp, S.T., Kelly, P.A., Battiste, V., and Dunbar, S. Types of tracking errors induced by concurrent secondary manual task. Proceedings of the 20th Annual Conference on Manual Control (pp. 299-304). Moffett Field, CA: Ames Research Center; 1984.

Klapp, S.T., Kelly, P.A., and Netick, A. Hesitations in continuous tracking induced by a concurrent discrete task. Human Factors. 29(3), 327-337; 1987.

Krol, J.P. Variations in ATC-workload as a function of variations in cockpit workload. Ergonomics. 14, 585-590; 1971.

Logan, G.D. On the use of a concurrent memory load to measure attention and automaticity. Journal of Experimental Psychology: Human Perception and Performance. 5, 189-207; 1970

Lysaght, R.J., Hill, S.G., Dick, A.O., Plamondon, B.D., Linton, P.M., Wierwille, W.W., Zaklad, A.L., Bittner, A.C., and Wherry, R.J. Operator workload: comprehensive review and evaluation of operator workload methodologies (Technical Report 851). Alexandria, VA: Army Research Institute for the Behavioral and Social Sciences; June 1989.

Schvaneveldt, R.W. (1969). Effects of complexity in simultaneous reaction time tasks. Journal of Experimental Psychology. 81, 289-296; 1969.

Smith, M.C. Effect of varying channel capacity on stimulus detection and discrimination. Journal of Experimental Psychology. 82, 520-526; 1969.

Wempe, T.E. and Baty, D.L. Human information processing rates during certain multiaxis tracking tasks with a concurrent auditory task. IEEE Transactions on Man-Machine Systems. 9, 129-138; 1968.

3.1.9.3 Classification Secondary Task

General description - "The subject must judge whether symbol pairs are identical in form. For example, to match letters either on a physical level (AA) or on a name level (Aa)" (Lysaght, et al., 1989, p. 233), or property (pepper is hot), or superset relation (an apple is a fruit). Cogntive processing requirements are discussed in Miller (1975).

Strengths and limitations - "Depending upon the requirements of the matching task, the task can impose demands on perceptual processes (physical match) and/or cognitive processes (name match or category match)" (Lysaght, et al., 1989, p. 233).

Damos (1985) reported that percentage correct scores for both single and dual-task performance were affected only by trial and not by either behavior pattern or pacing condition. Correct RT scores, however, were significantly related to trial and pacing by behavior pattern in the single-task condition and trial, behavior pattern, trial by pacing, and trial by behavior pattern in the dual-task condition.

Carter, Krause, and Harbeson (1986) reported that RT increased as the number of memory steps to verify a sentence increased. Slope was not a reliable measure of performance.

Beer, Gallaway, and Previc (1996) reported that performance on an aircraft classification task in singular task mode did not predict performance in dual-task mode.

Data requirements - The following data are used to assess performance of this task: mean RT for physical match, mean RT for category match, number of errors for physical match, and number of errors for category match (Lysaght, et al., 1989, p. 236).

Thresholds - Kobus, Russotti, Schlichting, Haskell, Carpenter, and Wojtowicz (1986) reported the following times to correctly classify one of five targets: visual = 224.6s, auditory = 189.6s, and multimodal (i.e., both visual and auditory) = 212.7s. These conditions were not significantly different.

Sources -

Beer, M.A. Gallaway, R.A., and Previc, R.H. Do individuals' visual recognition thresholds predict performance on concurrent attitude control flight tasks? The International Journal of Aviation Psychology. 6(3), 273-297, 1996.

Carter, R.C., Krause, M., and Harbeson, M.M. Beware the reliability of slope scores for individuals. Human Factors. 28(6): 673-683; 1986.

Damos, D. The relation between the Type A behavior pattern, pacing, and subjective workload under single- and dual-task conditions. Human Factors. 27(6), 675-680; 1985.

Kobus, D.A., Russotti, J., Schlichting, C., Haskell, G., Carpenter, S., and Wojtowicz, J. Multimodal detection and recognition performance of sonar operations. Human Factors. 28(1), 23-29; 1986.

Lysaght, R.J., Hill, S.G., Dick, A.O., Plamondon, B.D., Linton, P.M., Wierwille, W.W., Zaklad, A.L., Bittner, A.C., and Wherry, R.J. Operator workload: comprehensive review and evaluation of operator workload methodologies (Technical Report 851). Alexandria, VA: Army Research Institute for the Behavioral and Social Sciences; June 1989.

Miller, K. Processing capacity requirements for stimulus encoding. Acta Psychologica. 39, 393-410; 1975.

3.1.9.4 Cross-Adaptive Loading Secondary Task

General description - Cross-adaptive loading tasks are secondary tasks that the subject must perform only while primary-task performance meets or exceeds a previously established performance criterion (Kelly and Wargo, 1967).

Strengths and limitations - Cross-adaptive loading tasks are less likely to degrade performance on the primary task but are intrusive and, as such, difficult to use in nonlaboratory settings.

Data requirements - A well-defined, quantifiable criterion for primary-task performance as well as a method of monitoring this performance and cueing the subject when to perform the cross-adaptive loading task are all required.

Thresholds - Dependent on type of primary and cross-adaptive loading tasks being used.

Source -

Kelly, C.R. and Wargo, M.J. Crossadaptive operator loading tasks. Human Factors. 9, 395-404; 1967.

3.1.9.5 Detection Secondary Task

General description - "The subject must detect a specific stimulus or event which may or may not be presented with alternative events. For example, to detect which of 4 lights is flickering. The subject is usually alerted by a warning signal (e.g., tone) before the occurrence of such events, therefore attention is required intermittently." (Lysaght, et al., 1989, p. 233).

Strengths and limitations - "Such tasks are thought to impose demands on perceptual processes" (Lysaght, et al., 1989, p. 233).

Based on a review of five studies in which detection was a secondary task, Lysaght, et al. (1989) reported that, for dual-task pairings: performance of a primary classification task remained stable; and performance of tracking, memory, monitoring, and detection primary tasks degraded. In all cases, performance of the secondary detection task degraded (see Table 5).

TABLE 5.
References Listed by the Effect on Performance of Primary Tasks Paired with a Secondary Detection Task

Type	PRIMARY TASK			SECONDARY TASK		
	Stable	Degraded	Enhanced	Stable	Degraded	Enhanced
Detection		Wickens, Mountford, and Schreiner (1981)			Wickens, *et al.* (1981)	
Memory					Shulman and Greenberg (1971)	
Tracking		Wickens, *et al.* (1981)			Wickens, *et al.* (1981)	

from Lysaght, et al. (1989) p. 246

Data requirements - The following data are calculated for this task: mean RT for correct detections and number of correct detections.

Thresholds - Not stated.

Source -

Lysaght, R.J., Hill, S.G., Dick, A.O., Plamondon, B.D., Linton, P.M., Wierwille, W.W., Zaklad, A.L., Bittner, A.C., and Wherry, R.J. Operator workload: comprehensive review and evaluation of operator workload methodologies (Technical Report 851). Alexandria, VA: Army Research Institute for the Behavioral and Social Sciences; June 1989.

Shulman, H.G. and Greenberg, S.N. Perceptual deficit due to division of attention between memory and perception. Journal of Experimental Psychology, 88, 171-176; 1971.

Wickens, C.D., Mountford, S.J., and Schreiner, W. Multiple resources, task-hemispheric integrity, and individual differences in time sharing. Human Factors. 23, 211-229; 1981.

3.1.9.6 Distraction Secondary Task

General description - "The subject performs a task which is executed in a fairly automatic way such as counting aloud" (Lysaght, et al., 1989, p. 233).

Strengths and limitations - "Such a task is intended to distract the subject in order to prevent the rehearsal of information that may be needed for the primary task" (Lysaght, et al., 1989, p. 233).

Based on one study, Lysaght, et al. (1989) reported degraded performance on a memory primary task when paired with a distraction secondary task. Drory (1985) reported significantly shorter brake RTs and fewer steering wheel reversals when a secondary distraction task (i.e., state the last two digits of the current odometer reading) was paired with a basic driving task in a simulator. There were no effects of the secondary distraction task on tracking error, number of brake responses, or control light response.

Zeitlin (1995) used two auditory secondary tasks (delayed digit recall and random digit generation) while driving on road. Performance on both tasks degraded as traffic density and average speed increased.

Data requirements - Not stated.

Thresholds - Not stated.

Sources -

Drory, A. Effects of rest and secondary task on simulated truck-driving task performance. Human Factors. 27(2), 201-207; 1985.

Lysaght, R.J., Hill, S.G., Dick, A.O., Plamondon, B.D., Linton, P.M., Wierwille, W.W., Zaklad, A.L., Bittner, A.C., and Wherry, R.J. Operator workload: comprehensive review and evaluation of operator workload methodologies (Technical Report 851). Alexandria, VA: Army Research Institute for the Behavioral and Social Sciences; June 1989.

Zeitlin, L.R. Estimates of driver mental workload: a long-term field trial of two subsidiary tasks. Human Factors. 37(3), 611-621; 1995.

3.1.9.7 Driving Secondary Task

General description - "The subject operates a driving simulator or actual motor vehicle" (Lysaght, et al., 1989, p. 232).

Strengths and limitations - This "task involves complex psychomotor skills" (Lysaght, et al., 1989, p. 232). Johnson and Haygood (1984) varied the difficulty of a primary simulated driving task by varying the road width. The secondary task was a visual choice RT task. Tracking score was highest when the difficulty of the primary task was adapted as a function of primary task performance. It was lowest when the difficulty was fixed.

Brouwer, Waterink, van Wolffelaar, and Rothengatten (1991) reported that older adults (mean age 64.4) were significantly worse than younger adults (mean age 26.1) in dual task performance of compensatory lane tracking with a timed, self-paced visual analysis task.

Korteling (1994) did not find a significant difference in steering performance between single task (steering) versus dual task (addition of car following task) between young (21 to 34) and old (65 to 74 year old) drivers. There was, however, 24% performance deterioration in car-following performance with the addition of a steering task.

Data requirements - The experimenter should be able to record: total time to complete a trial, number of acceleration rate changes, number of gear changes, number of footbrake operations, number of steering reversals, number of obstacles hit, high pass steering deviation, yaw deviation, and lateral deviation" (Lysaght, et al., 1989, p. 235).

Thresholds - Not stated.

Sources -

Brouwer, W.H., Waterink, W., van Wolffelaar, P.C., and Rothengatten, T. Divided attention in experienced young and older drivers: lane tracking and visual analysis in a dynamic driving simulator. Human Factors. 33(5), 573-582; 1991.

Johnson, D.F., and Haygood, R.C. The use of secondary tasks in adaptive training. Human Factors. 26(1), 105-108; 1984.

Korteling, J.E. Effects of aging, skill modification, and demand alternation on multiple-task performance. Human Factors. 36(1), 27-43; 1994.

Lysaght, R.J., Hill, S.G., Dick, A.O., Plamondon, B.D., Linton, P.M., Wierwille, W.W., Zaklad, A.L., Bittner, A.C., and Wherry, R.J. Operator workload: comprehensive review and evalaution of operator workload methodologies (Technical Report 851). Alexandria, VA: Army Research Institute for the Behavioral and Social Sciences; June 1989.

3.1.9.8 Identification/Shadowing Secondary Task

General description - "The subject identifies changing symbols (digits and/or letters) that appear on a visual display by writing or verbalizing, or repeating a spoken passage as it occurs" (Lysaght, et al., 1989, p. 233).

Strengths and limitations - "Such tasks are thought to impose demands on perceptual processes (i.e., attention)" (Lysaght, et al., 1989, p. 233). Wierwille and Connor (1983), however, reported that a digit shadowing task was not sensitive to variations in workload. Their task was control on moving-base aircraft simulator. Workload was varied by manipulating pitch-stability level, wind-gust disturbance level, and crosswind velocity and direction.

Savage, Wierwille, and Cordes (1978) evaluated the sensitivity of four dependent measures to workload manipulations (1, 2, 3, or 4 meters) to a primary monitoring task. The number of random digits spoken on the secondary task was the most sensitive to workload. The longest consecutive string of spoken digits and the number of triplets spoken were also significantly affected by workload. The longest interval between spoken responses, however, was not sensitive to workload manipulations of the primary task.

Based on nine studies with an identification secondary task, Lysaght, et al. (1989) reported that performance: remained stable for an identification primary task; and degraded for tracking, memory, detection, driving, and spatial transformation primary tasks. Performance of the identification secondary task: remained stable for tracking and identification primary tasks; and degraded for monitoring, detection, driving, and spatial transformation primary tasks (see Table 6).

TABLE 6.
References Listed by the Effect on Performance of Primary Tasks Paired with a Secondary Identification Task

Type	PRIMARY TASK			SECONDARY TASK		
	Stable	Degraded	Enhanced	Stable	Degraded	Enhanced
Detection		Price (1975)			Price (1975)	
Driving		Hicks and Wierwille (1979)			Wierwille, Gutmann, Hicks, and Muto (1977)	
Identification	Allport, Antonis, and Reynolds (1972)			Allport, *et al.* (1972)		
Memory		Mitsuda (1968)				
Monitoring					Savage, Wierwille and Cordes (1978)	
Spatial Transformation		Fournier and Stager (1976)			Fournier and Stager (1976)	
Tracking		Gabay and Merhav (1977)		Gabay and Merhav (1977)		

from Lysaght, et al. (1989) p. 246

Data requirements - The following data are used for this task: number of words correct/minute, number of digits spoken, mean time interval between spoken digits, and number of errors of omission (Lysaght, et al., 1989, p. 236).

Thresholds - Not stated.

Sources -

Allport, D.A., Antonis, B., and Reynolds, P. On the division of attention: A disproof of the single channel hypothesis. Quarterly Journal of Experimental Psychology. 24, 225-235; 1972.

Fournier, B.A. and Stager, P. Concurrent validation of a dual-task selection test. Journal of Applied Psychology. 5, 589-595; 1976.

Gabay, E. and Merhav, S.J. Identification of a parametric model of the human operator in closed-loop control tasks. IEEE Transactions on Systems, Man, and Cybernetics. SMC-7, 284-292; 1977.

Hicks, T.G. and Wierwille, W.W. Comparison of five mental workload assessment procedures in a moving-base driving simulator. Human Factors. 21, 129-142; 1979.

Lysaght, R.J., Hill, S.G., Dick, A.O., Plamondon, B.D., Linton, P.M., Wierwille, W.W., Zaklad, A.L., Bittner, A.C., and Wherry, R.J. Operator workload: comprehensive review and evaluation of operator workload methodologies (Technical Report 851). Alexandria, VA: Army Research Institute for the Behavioral and Social Sciences; June 1989.

Mitsuda, M. Effects of a subsidiary task on backward recall. Journal of Verbal Learning and Verbal Behavior. 7, 722-725; 1968.

Price, D.L. The effects of certain gimbal orders on target acquisition and workload. Human Factors. 20, 649-654; 1975.

Savage, R.E., Wierwille, W.W., and Cordes, R.E. Evaluating the sensitivity of various measures of operator workload using random digits as a secondary task. Human Factors. 20, 649-654; 1978.

Wierwille, W.W. and Connor, S.A. Evaluation of 20 workload measures using a psychomotor task in a moving-base aircraft simulator. Human Factors. 25(1), 1-16, 1983.

Wierwille, W.W., Gutmann, J.C., Hicks, T.G., and Muto, W.H. Secondary task measurement of workload as a function of simulated vehicle dynamics and driving conditions. Human Factors. 19, 557-565; 1977.

3.1.9.9 Lexical Decision Secondary Task

General description - "Typically, the subject is briefly presented with a sequence of letters and must judge whether this letter sequence forms a word or a non-word" (Lysaght, et al., 1989, p. 233).

Strengths and limitations - "This task is thought to impose heavy demands on semantic memory processes" (Lysaght, et al., 1989, p. 233).

Data requirements - Mean RT for correct responses is used as data for this task.

Thresholds - Not stated.

Source -

Lysaght, R.J., Hill, S.G., Dick, A.O., Plamondon, B.D., Linton, P.M., Wierwille, W.W., Zaklad, A.L., Bittner, A.C., and Wherry, R.J. Operator workload: comprehensive review and evaluation of operator workload methodologies (Technical Report 851). Alexandria, VA: Army Research Institute for the Behavioral and Social Sciences; June 1989.

3.1.9.10 Memory-Scanning Secondary Task

General description - These secondary tasks require a subject to memorize a list of letters, numbers, and/or shapes and then indicate whether a probe stimulus is a member of that set. Typically, there is a linear relation between the number of items in the memorized list and RT to the probe stimulus.

Strengths and limitations - The slope of the linear function may reflect memory-scanning rate. Fisk and Hodge (1992) reported no significant differences in RT in a single task performance of a memory-scanning task after 32 days without practice. But Carter, Krause, and Harbeson (1986) warned that the slope may be less reliable than the RTs used to calculate the slope.

These tasks may not produce good estimates of memory load, since: (1) addition of memory load does not affect the slope, and (2) the slope is affected by stimulus-response compatibility effects. Further, Wierwille and Connor (1983), however, reported that a memory-scanning task was not sensitive to workload. Their primary task was control a moving-base aircraft simulator. Workload was varied by manipulating pitch-stability level, wind-gust disturbance level, and crosswind direction and velocity. However, Park and Lee (1992) reported memory tasks significantly predicted flight performance of pilot trainees.

Based on twenty-five studies using a memory secondary task, Lysaght, et al. (1989) reported that performance: remained stable on tracking, mental math, monitoring, problem solving, and driving primary tasks; degraded on tracking, choice RT, memory, monitoring, problem solving, detection, identification, classification, and distraction primary tasks; and improved on a tracking primary task. Performance of the memory secondary task: remained stable with a tracking primary task; and degraded when paired with tracking, choice RT, memory, mental math, monitoring, detection, identification, classification, and driving primary tasks (see Table 7).

Data requirements - RT to items from the memorized list must be at asymptote to ensure that no additional learning will take place during the experiment. The presentation of the probe stimulus is intrusive and, thus, may be difficult to use in nonlaboratory settings.

Thresholds - 40 msec for RT.

Sources -

Allport, D.A., Antonis, B., and Reynolds, P. On the division of attention: A disproof of the single channel hypothesis. Quarterly Journal of Experimental Psychology. 24, 225-235; 1972.

Broadbent, D.E. and Gregory, M. On the interaction of S-R compatibility with other variables affecting reaction time. British Journal of Psychology. 56, 61-67; 1965.

Broadbent, D.E. and Heron, A. Effects of a subsidiary task on performance involving immediate memory by younger and older men. British Journal of Psychology. 53, 189-198; 1962.

Brown, I.D. Measuring the "spare mental capacity" of car drivers by a subsidiary auditory task. Ergonomics. 5, 247-250; 1962.

Brown, I.D. A comparison of two subsidiary tasks used to measure fatigue in car drivers. Ergonomics. 8, 467-473; 1965.

Brown, I.D. Subjective and objective comparisons of successful and unsuccessful trainee drivers. Ergonomics. 9, 49-56; 1966.

TABLE 7.
References Listed by the Effect on Performance of Primary Tasks Paired with a Secondary Memory Task

Type	PRIMARY TASK			SECONDARY TASK		
	Stable	Degraded	Enhanced	Stable	Degraded	Enhanced
Choice RT		Broadbent and Gregory (1965) Keele and Boies (1973)			Broadbent and Gregory (1965)	
Classification		Wickens, *et al.* (1981)			Wickens, *et al.* (1981)	
Detection		Wickens, *et al.* (1981)			Wickens, *et al.* (1981)	
Distraction		Broadbent and Heron (1962)				
Driving	Brown (1962, 1965, 1966) Brown and Poulton (1961) Wetherell (1981)			Brown (1965)	Brown (1962, 1965, 1966) Brown and Poulton (1961) Wetherell (1981)	
Identification		Klein (1976)			Allpor, Antonis, and Reynolds (1972)	
Memory		Broadbent and Heron (1962) Chow and Murdock (1975)		Shulman and Greenberg (1971)		
Mental Math	Mandler and Worden (1973)			Mandler and Worden (1973)		
Monitoring	Chechile, Butler, Gutowski, and Palmer (1979) Moskowitz and McGlothlin (1974)	Chiles and Alluisi (1979)		Chechile, *et al.* (1979) Chiles and Alluisi (1979) Mandler and Worden (1973) Moskowitz and McGlothlin (1974)		
Problem Solving	Daniel, Florek, Kosinar, and Strizenec (1969)	Stager and Zufelt (1972)				
Tracking	Finkelman and Glass (1970) Zeitlin and Finkelman (1975)	Heimstra (1970) Huddleston and Wilson (1971) Noble, Trumbo, and Fowler (1967) Trumbo and Milone (1971) Wickens and Kessel (1980) Wickens, Mountford, and Schreiner (1981)	Tsang and Wickens (1984)	Noble, *et al.* (1967) Trumbo and Milone (1971)	Finkelman and Glass (1970) Heimstra (1970) Huddleston and Wilson (1971) Tsang and Wickens (1984) Wickens and Kessel (1980) Wickens, *et al.* (1981)	

from Lysaght, et al. (1989) p. 246

Brown, I.D. and Poulton, E.C. Measuring the spare "mental capacity" of car drivers by a subsidiary task. Ergonomics. 4, 35-40; 1961.

Carter, R.C., Krause, M., and Harbeson, M.M. Beware the reliability of slope scores for individuals. Human Factors. 28(6): 673-683; 1986.

Chechile, R.A., Butler, K., Gutowski, W., and Palmer, E.A. Division of attention as a function of the number of steps, visual shifts and memory load. Proceedings of the 15th Annual Conference on Manual Control (pp. 71-81). Dayton, OH: Wright State University; 1979.

Chiles, W.D. and Alluisi, E.A. On the specification of operator or occupa-tional workload performance-measurement methods. Human Factors. 21, 515-528; 1979.

Chow, S.L. and Murdock, B.B. The effect of a subsidiary task on iconic memory. Memory and Cognition. 3, 678-688; 1975.

Daniel, J., Florek, H., Kosinar, V., and Strizenec, M. Investigation of an operator's characteristics by means of factorial analysis. Studia Psychologica. 11, 10-22; 1969.

Finkelman, J.M. and Glass, D.C. Reappraisal of the relationship between noise and human performance by means of a subsidiary task measure. Journal of Applied Psychology. 54, 211-213; 1970.

Fisk, A.D. and Hodge, K.A. Retention of trained performance in consistent mapping search after extended delay. Human Factors. 34(2): 147-164; 1992.

Heimstra, N.W. The effects of "stress fatigue" on performance in a simulated driving situation. Ergonomics. 13, 209-218; 1970.

Huddleston, J.H.F. and Wilson, R.V. An evaluation of the usefulness of four secondary tasks in assessing the effect of a lag in simulated aircraft dynamics. Ergonomics. 14, 371-380; 1971.

Keele, S.W. and Boies, S.J. Processing demands of sequential information. Memory and Cognition. 1, 85-90; 1973.

Klein, G.A. Effect of attentional demands on context utilization. Journal of Educational Psychology. 68, 25-31; 1976.

Lysaght, R.J., Hill, S.G., Dick, A.O., Plamondon, B.D., Linton, P.M., Wierwille, W.W., Zaklad, A.L., Bittner, A.C., and Wherry, R.J. Operator workload: comprehensive review and evaluation of operator workload methodologies (Technical Report 851). Alexandria, VA: Army Research Institute for the Behavioral and Social Sciences; June 1989.

Mandler, G. and Worden, P.E. Semantic processing without permanent storage. Journal of Experimental Psychology. 100, 277-283; 1973.

Moskowitz, H. and McGlothlin, W. Effects of marijuana on auditory signal detection. Psychopharmacologia. 40, 137-145; 1974.

Noble, M., Trumbo, D., and Fowler, F. Further evidence on secondary task interference in tracking. Journal of Experimental Psychology. 73, 146-149; 1967.

Park, K.S. and Lee, S.W. A computer-aided aptitude test for predicting flight performance of trainees. Human Factors. 34(2), 189-204; 1992.

Shulman, H.G. and Greenberg, S.N. Perceptual deficit due to division of attention between memory and perception. Journal of Experimental Psychology. 88, 171-176; 1971.

Stager, P. and Zufelt, K. Dual-task method in determining load differences. Journal of Experimental Psychology. 94, 113-115; 1972.

Trumbo, D. and Milone, F. Primary task performance as a function of encoding, retention, and recall in a secondary task. Journal of Experimental Psychology. 91, 273-279; 1971.

Tsang, P.S. and Wickens, C.D. The effects of task structures on time-sharing efficiency and resource allocation optimality. Proceedings of the 20th Annual Conference on Manual Control (pp. 305-317). Moffett Field, CA: Ames Research Center; 1984.

Wetherell, A. The efficacy of some auditory-vocal subsidiary tasks as measures of the mental load on male and female drivers. Ergonomics. 24, 197-214; 1981.

Wickens, C.D. and Kessel, C. Processing resource demands of failure detection in dynamic systems. Journal of Experimental Psychology: Human Perception and Performance. 6, 564-577; 1980.

Wickens, C.D., Mountford, S.J., and Schreiner, W. Multiple resources, task-hemispheric integrity, and individual differences in time sharing. Human Factors. 23, 211-229; 1981.

Wierwille, W.W. and Connor, S.A. Evaluation of 20 workload measures using a psychomotor task in a moving-base aircraft simulator. Human Factors. 25(1), 1-16, 1983.

Zeitlin, L.R. and Finkelman, J.M. Research note: Subsidiary task techniques of digit generation and digit recall indirect measures of operator loading. Human Factors. 17, 218-220; 1975.

3.1.9.11 Mental Mathematics Secondary Task

General description - Subjects are asked to perform arithmetic operations (i.e., addition, subtraction, multiplication, and division) on sets of visually or aurally presented digits.

Strengths and limitations - The major strength of this workload measure is its ability to discriminate between good and poor operators and high and low workload. For example, Ramacci and Rota (1975) required pilot applicants to perform progressive subtraction during their initial flight training. They reported that the number of subtractions performed increased while the percent of errors decreased with flight experience. Further, successful applicants performed more subtractions and had a lower percentage of errors than those applicants who were not accepted.

Green and Flux (1977) required pilots to add the digit three to aurally presented digits during a simulated flight. They reported increased performance time of the secondary task as the workload associated with the primary task increased. Huddleston and Wilson (1971) asked pilots to determine if digits were odd or even, their sum was odd or even, two consecutive digits were the same or different, or every other digit was the same or different. Again, secondary task performance discriminated between high and low workload on the primary task. The major disadvantage of secondary tasks is their intrusion into the primary task. Harms (1986) reported similar results for a driving task. However, Andre, Heers, and Cashion (1995) reported greater rmse in roll, pitch, and yaw in a primary simulated flight task when paired with a mental mathematics secondary task (i.e., fuel range).

Mental mathematics tasks have also been used in the laboratory. For example, Kramer, Wickens, and Donchin (1984) reported a significant increase in tracking error on the primary task when the secondary task was counting flashes. In addition, Damos

(1985) required subjects to calculate the absolute difference between the digit currently presented visually and the digit that had preceded it. In the single-task condition, percentage correct scores were significantly related to trial. Correct RT scores were related to trial and trial by pacing condition. In the dual-task condition, percentage correct scores were not significantly related to trial, behavior pattern, or pacing condition. Correct RT in the dual-task condition, however, was related to trial, trial by pacing, and trial by behavior pattern.

Based on fifteen studies that used a mental math secondary task, Lysaght, et al. (1989) reported that performance remained the same for tracking, driving, and tapping primary tasks and degraded for tracking, choice RT, memory, monitoring, simple RT, and detection primary tasks. Performance of the mental math secondary task remained stable with a tracking primary task; degraded with tracking, choice RT, monitoring, detection, driving, and tapping primary tasks; and improved with tracking primary task (see Table 8).

Data requirements - The following data are calculated: number of correct responses, mean RT for correct responses, and number of incorrect responses (Lysaght, et al., 1989, p. 235). The researcher should compare primary task performance with and without a secondary task to ensure that pilots are not sacrificing primary task performance to enhance secondary task performance.

Thresholds - Not stated.

Sources -

Andre, A.D., Heers, S.T., and Cashion, P.A. Effects of workload preview on task scheduling during simulated instrument flight. International Journal of Aviation Psychology. 5(1), 5-23, 1995.

Bahrick, H.P., Noble, M., and Fitts, P.M. Extra-task performance as a measure of learning task. Journal of Experimental Psychology. 4, 299-302; 1954.

Brown, I.D. and Poulton, E.C. Measuring the spare "mental capacity" of car drivers by a subsidiary task. Ergonomics. 4, 35-40; 1961.

Chiles, W.D. and Jennings, A.E. Effects of alcohol on complex performance. Human Factors. 12, 605-612; 1970.

Damos, D. The relation between the type A behavior pattern, pacing, and subjective workload under single- and dual-task conditions. Human Factors. 27(6), 675-680; 1985.

Fisher, S. The microstructure of dual task interaction. 1. The patterning of main task response within secondary-task intervals. Perception. 4, 267-290; 1975.

Green, R. and Flux, R. Auditory communication and workload. Proceedings of NATO Advisory Group for Aerospace Research and Development Conference on Methods to Assess Workload, AGARD-CPP-216, A4-1-A4-8; 1977.

Harms, L. Drivers' attentional response to environmental variations: a dual-task real traffic study. In A.G. Gale, M.H. Freeman, C.M. Haslegrave, P. Smith, and S.P. Taylor (Eds.) Vision in vehicles (pp. 131-138). Amsterdam: North Holland; 1986.

Heimstra, N.W. The effects of "stress fatigue" on performance in a simulated driving situation. Ergonomics. 13, 209-218; 1970.

Huddleston, J.H.F. and Wilson, R.V. An evaluation of the usefulness of four secondary tasks in assessing the effect of a log in simulated aircraft dynamics. Ergonomics. 14, 371-380; 1971.

TABLE 8.
References Listed by the Effect on Performance of Primary Tasks Paired with a Secondary Task

Type	PRIMARY TASK			SECONDARY TASK		
	Stable	Degraded	Enhanced	Stable	Degraded	Enhanced
Choice RT		Chiles and Jennings (1970) Fisher (1975) Keele (1967)			Fisher (1975) Keele (1967) Schouten, Kalsbeek, and Leopold (1962)	
Detection		Jaschinski (1982)			Jaschinski (1982)	
Driving	Brown and Poulton (1961) Wetherell (1981)				Brown and Poulton (1961) Wetherell (1981)	
Memory		Roediger, Knight, and Kantowitz (1977) Silverstein and Glanzer (1971)				
Monitoring		Chiles and Jennings (1970) Kahneman, Beatty, and Pollack (1967)			Chiles and Jennings (1970) Kahneman, *et al.* (1967)	
Simple RT		Chiles and Jennings (1970) Green and Flux (1977)* Wierwille and Connor (1983)*			Green and Flux (1977)*	
Simulated Flight Task	Green and Flux (1977)* Wierwille and Connor (1983)*	Andre, Heers, and Cashion (1995)*			Green and Flux (1977)*	
Tapping	Kantowitz and Knight (1974)				Kantowitz and Knight (1974, 1976)	
Tracking	Huddleston and Wilson (1971)	Bahrick, Noble, and Fitts (1954) Chiles and Jennings (1970) Heimstra (1970) McLeod (1973) Wickens, Mountford, and Schreiner (1981)	Kramer, Wickens, and Donchin (1984)	Bahrick, *et al.* (1954) Heimstra (1970)	Huddleston and Wilson (1971) McLeod (1973) Wickens, *et al.* (1981)	Chiles and Jennings (1970)

from Lysaght, et al. (1989) p. 247
**Not included in Lysaght, et al. (1989)*

Jaschinski, W. Conditions of emergency lighting. Ergonomics. 25, 363-372; 1982.

Kahneman, D., Beatty, J., and Pollack, I. Perceptual deficit during a mental task. Science. 157, 218-219; 1967.

Kantowitz, B.H. and Knight, J.L. Testing tapping time-sharing. Journal of Experimental Psychology. 103, 331-336; 1974.

Kantowitz, B.H. and Knight, J.L. Testing tapping time sharing: II. Auditory secondary task. Acta Psychologica. 40, 343-362; 1976.

Keele, S.W. Compatibility and time-sharing in serial reaction time. Journal of Experimental Psychology. 75, 529-539; 1967.

Kramer, A.F., Wickens, C.D., and Donchin, E. Performance and enhancements under dual-task conditions. Annual Conference on Manual Control, June 1984, 21-35.

Lysaght, R.J., Hill, S.G., Dick, A.O., Plamondon, B.D., Linton, P.M., Wierwille, W.W., Zaklad, A.L., Bittner, A.C., and Wherry, R.J. Operator workload: comprehensive review and evaluation of operator workload methodologies (Technical Report 851). Alexandria, VA: Army Research Institute for the Behavioral and Social Sciences; June 1989.

McLeod, P.D. Interference of "attend to and learn" tasks with tracking. Journal of Experimental Psychology. 99, 330-333; 1973.

Ramacci, C.A. and Rota, P. Flight fitness and psycho-physiological behavior of applicant pilots in the first flight missions. Proceedings of NATO Advisory Group for Aerospace Research and Development. 153, B8; 1975.

Roediger, H.L., Knight, J.L., and Kantowitz, B.H. Inferring delay in short-term memory: The issue of capacity. Memory and Cognition. 5, 167-176; 1977.

Schouten, J.F., Kalsbeek, J.W.H., and Leopold, F.F. On the evaluation of perceptual and mental load. Ergonomics. 5, 251-260; 1962.

Silverstein, C. and Glanzer, M. Concurrent task in free recall: Differential effects of LTS and STS. Psychonomic Science. 22, 367-368; 1971.

Wetherell, A. The efficacy of some auditory-vocal subsidiary tasks as measures of the mental load on male and female drivers. Ergonomics. 24, 197-214; 1981.

Wickens, C.D., Mountford, S.J., and Schreiner, W. Multiple resources, task-hemispheric integrity, and individual differences in time-sharing. Human Factors. 23, 211-229; 1981.

Wierwille, W.W. and Connor, S. Evaluation of 20 workload measures using a psychomotor task in a moving base aircraft simulator. Human Factors, 25, 1-16; 1983.

3.1.9.12 Michon Interval Production Secondary Task

General description - "The Michon paradigm of interval production requires the subject to generate a series of regular time intervals by executing a motor response (i.e., a single finger tap [every] 2 sec.). No sensory input is required." (Lysaght, et al., 1989, p. 233). Based on a review of six studies in which the Michon Interval Production task was the secondary task, Lysaght, et al. (1989) stated that, in dual-task pairings, performance of flight simulation and driving primary tasks remained stable; performance of monitoring, problem solving, detection, psychomotor, Sternberg, tracking, choice RT, memory, and mental math primary tasks degraded; and performance of simple RT primary task improved. In these same pairings, performance of the Michon Interval Production task remained stable with monitoring, Sternberg, flight simulation, and memory primary tasks; and degraded with problem solving, simple RT, detection, psychomotor, flight simulation, driving, tracking, choice RT, and mental math primary tasks (see Table 9).

TABLE 9.
References Listed by the Effect on Performance of Primary Tasks Paired with a Secondary Michon Interval Production Task

Type	PRIMARY TASK			SECONDARY TASK		
	Stable	Degraded	Enhanced	Stable	Degraded	Enhanced
Choice RT		Michon (1964)			Michon (1964)	
Detection		Michon (1964)			Michon (1964)	
Driving	Brown (1967)	Brown, Simmonds, and Tickner (1967)*			Brown (1967) Brown, *et al.* (1967)*	
Flight Simulation	Wierwille, Casali, Connor, and Rahimi (1985)			Wierwille, Casali, Connor, and Rahimi (1985)	Wierwille, *et al.* (1985)	
Memory		Roediger, Knight, and Kantowitz (1977)		Roediger, Knight, and Kantowitz (1977)		
Mental Math		Michon (1964)			Michon (1964)	
Monitoring		Shingledecker, Actonand Crabtree (1983)		Shingledecker, *et al.* (1983)		
Problem Solving		Michon (1964)			Michon (1964)	
Psychomotor		Michon (1964)			Michon (1964)	
Simple RT			Vroon (1973)		Vroon (1973)	
Sternberg		Shingledecker, *et al.* (1983)		Shingledecker, *et al.* (1983)		
Tracking		Shingledecker, *et al.* (1983)			Shingledecker, *et al.* (1983)	

from Lysaght, et al. (1989) p. 245
**Not included in Lysaght, et al. (1989)*

Strengths and limitations - "This task is thought to impose heavy demand on motor output/response resources. It has been demonstrated with high demand primary tasks that subjects exhibit irregular or variable tapping rates." (Lysaght, et al., 1989, p. 233). Crabtree, Bateman, and Acton (1984) reported that scores on this secondary task discriminated the workload of three switch-setting tasks. Johannsen, Pfendler, and Stein (1976) reported similar results for autopilot evaluation in a fixed-based flight simulator. Wierwille, Rahimi, and Casali (1985), however, reported that tapping regularity was not affected by variations in the difficulty of a mathematical problem to be solved during simulated flight.

Data requirements - The following data are calculated: mean interval per trial, standard deviation of interval per trial, and sum of differences between successive intervals per minute of total time (Lysaght, et al., 1989, p. 235).

Thresholds - Not stated.

Sources -

Brown, I.D. Measurement of control skills, vigilance, and performance on a subsidiary task during twelve hours of car driving. Ergonomics. 10, 665-673; 1967.

Brown, I.D., ., D.C.V., and Tickner, A.H. Measurement of control skills, vigilance, and performance on a subsidiary task during 12 hours of car driving. Ergonomics, 10, 655-673; 1967.

Crabtree, M.S., Bateman, R.P., and Acton, W.H. Benefits of using objective and subjective workload measures. Proceedings of the 28th Annual Meeting of the Human Factors Society (pp. 950-953). Santa Monica, CA; Human Factors Society; 1984.

Johannsen, G., Pfendler, C., and Stein, W. Human performance and workload in simulated landing approaches with autopilot failures. In N. Moray (Ed.) Mental workload, its theory and measurement (pp. 101-104). New York: Plenum Press; 1976.

Lysaght, R.J., Hill, S.G., Dick, A.O., Plamondon, B.D., Linton, P.M., Wierwille, W.W., Zaklad, A.L., Bittner, A.C., and Wherry, R.J. Operator workload: comprehensive review and evaluation of operator workload methodologies (Technical Report 851). Alexandria, VA: Army Research Institute for the Behavioral and Social Sciences; June 1989.

Michon, J.A. A note on the measurement of perceptual motor load. Ergonomics. 7, 461-463; 1964.

Roediger, H.L., Knight, J.L., and Kantowitz, B.H. Inferring decay in short-term memory: The issue of capacity. Memory and Cognition. 5, 167-176; 1977.

Shingledecker, C.A., Acton, W., and Crabtree, M.S. Development and application of a criterion task set for workload metric evaluation (SAE Technical Paper No. 831419). Warrendale, PA: Society of Automotive Engineers; 1983.

Vroon, P.A. Tapping rate as a measure of expectancy in terms of response and attention limitation. Journal of Experimental Psychology. 101, 183-185; 1973.

Wierwille, W.W., Casali, J.G., Connor, S.A., and Rahimi, M. Evaluation of the sensitivity and intrusion of mental workload estimation techniques. In W. Roner (Ed.) Advances in man-machine systems research, Volume 2 (pp. 51-127). Greenwich, CT: J.A.I. Press; 1985.

Wierwille, W.W., Rahimi, M., and Casali, J.G. Evaluation of 16 measures of mental workload using a simulated flight task emphasizing mediational activity. Human Factors. 27(5), 489-502; 1985.

3.1.9.13 Monitoring Secondary Task

General description - Subjects are asked to respond either manually or verbally to the onset of visual or auditory stimuli. Both the time to respond and the accuracy of the response have been used as workload measures.

Strengths and limitations - The major advantage of the monitoring-task technique is its relevance to system safety. It is also able to discriminate among levels of automation and workload. For example, Anderson and Toivanen (1970) used a force-paced digit-naming task as a secondary task to investigate the effects of varying levels of automation in a helicopter simulator. Bortolussi, Kantowitz, and Hart (1986) reported significant differences in two- and four-choice visual RT in easy and difficult flight scenarios. Bortolussi, Hart, and Shively (1987) reported significantly longer RTs in a four-choice reaction-time task during a high-difficulty scenario than during a low-difficulty one. Brown (1969) studied it in relation to flicker.

Based on the results of thirty-six studies that included a secondary monitoring task, Lysaght, et al. (1989) reported that performance remained stable on tracking, choice RT, memory, mental math, problem solving, identification, and driving primary tasks; degraded on tracking, choice RT, memory, monitoring, detection, and driving primary tasks; and improved on a monitoring primary task. Performance of the monitoring secondary task remained stable when paired with tracking, memory, monitoring, flight simulation, and driving primary tasks; degraded when paired with tracking, choice RT, mental math, monitoring, problem solving, detection, identification and driving primary tasks; and improved when paired with tracking and driving primary tasks (see Table 10).

Data requirements - The experimenter should calculate: number of correct detections, number of incorrect detections, number of errors of omission, mean RT for correct detections, and mean RT for incorrect detections (Lysaght, et al., 1989, p. 235).

The Knowles (1963) guidelines are appropriate in selecting a vigilance task. In addition, the modality of the task must not interfere with performance of the primary task, for example, requiring a verbal response while a pilot is communicating with Air Traffic Control or other crewmembers.

Thresholds - Not stated.

Sources -

Anderson, P.A. and Toivanen, M.L. Effects of varying levels of autopilot assistance and workload on pilot performance in the helicopter formation flight mode (Technical Report JANAIR 680610). Washington, D.C.: Office of Naval Research, March 1970.

Bell, P.A. Effects of noise and heat stress on primary and subsidiary task performance. Human Factors. 20, 749-752; 1978.

Bergeron, H.P. Pilot response in combined control tasks. Human Factors. 10, 277-282; 1968.

Boggs, D.H. and Simon, J.R. Differential effect of noise on tasks of varying complexity. Journal Applied Psychology. 52, 148-153; 1968.

Bortolussi, M.R., Hart, S.G., and Shively, R.J. Measuring moment-to-moment pilot workload using synchronous presentations of secondary tasks in a motion-base trainer. Proceedings of the 4th Symposium on Aviation Psychology. 651-657; 1987.

Bortolussi, M.R., Kantowitz, B.H., and Hart, S.G. Measuring pilot workload in a motion base trainer: A comparison of four techniques. Applied Ergonomics. 17, 278-283; 1986.

Brown, I.D. Measuring the "spare mental capacity" of car drivers by a subsidiary auditory task. Ergonomics. 5, 247-250; 1962.

Brown, I.D. A comparison of two subsidiary tasks used to measure fatigue in car drivers. Ergonomics. 8, 467-473; 1965.

Brown, I.D. Measurement of control skills, vigilance, and performance on a subsidiary task during twelve hours of car driving. Ergonomics. 10, 665-673; 1967.

Brown, J.L. Flicker and intermittent stimulation. In C.H. Graham (Ed.) Vision and visual perception. New York: Wiley; 1969.

Chechile, R.A., Butler, K., Gutowski, W., and Palmer, E.A. Division of attention as a function of the number of steps, visual shifts, and memory load. Proceedings of the 15th Annual Conference on Manual Control (pp. 71-81). Dayton, OH: Wright State University; 1979.

Chiles, W.D., Jennings, A.E., and Alluisi, E.C. Measurement and scaling of workload in complex performance. Aviation, Space, and Environmental Medicine. 50, 376-381; 1979.

TABLE 10.
References Listed by the Effect on Performance of Primary Tasks Paired with a Secondary Monitoring Task

Type	PRIMARY TASK			SECONDARY TASK		
	Stable	Degraded	Enhanced	Stable	Degraded	Enhanced
Choice RT	Boggs and Simon (1968)	Hilgendorf (1967)			Hilgendorf (1967)	
Detection		Dewar, Ells, and Mundy (1976) Tyler and Halcomb (1974)			Tyler and Halcomb (1974)	
Driving	Brown (1962, 1967) Hoffman and Jorbert (1966) Wetherell (1981)	Brown (1965)		Hoffman and Jorbert (1966)	Brown (1962, 1965)	Brown (1967)
Flight Simulation				Soliday and Schohan (1965)		
Identification	Dornic (1980)				Dornic (1980) Chiles, Jennings, and Alluisi (1979)	
Memory	Tyler and Halcomb (1974)	Chow and Murdock (1975) Lindsay and Norman (1969) Mitsuda (1968)		Lindsay and Norman (1968)		
Mental Math	Dornic (1980)				Chiles, Jennings, and Alluisi (1979) Dornic (1980)	
Monitoring		Chechile, Butler, Gutowski, and Palmer (1979) Fleishman (1965) Goldstein and Dorfman (1978) Hohmuth (1970) Long (1976) Stager and Muter (1971)	McGrath (1965)	Stager and Muter (1971)	Chechile, *et al.* (1979) Hohmuth (1970) Long (1976)	
Problem Solving	Wright, Holloway, and Aldrich (1974)				Chiles, *et al.* (1979) Wright, *et al.* (1974)	
Tracking	Bell (1978) Figarola and Billings (1966) Gabriel and Burrows (1968) Huddleston and Wilson (1971) Kelley and Wargo (1967) Kyriakides and Leventhal (1967) Schori and Jones (1975)	Bergeron (1968) Heimstra (1970) Herman (1965) Kramer, Wickens and Donchin (1984) Malmstrom, Reed, and Randle (1983) Monty and Ruby (1965) Putz and Rothe (1974)		Figarola and Billings (1966) Kramer, *et al.* (1984) Malmstrom, *et al.* (1983)	Bell (1978) Bergeron (1968) Gabriel and Burrows (1968) Herman (1965) Huddleston and Wilson (1971) Kelley and Wargo (1967) Kyriakides and Leventhal (1967) Monty and Ruby (1965) Putz and Rothe (1974) Schori and Jones (1975)	Heimstra (1970)

from Lysaght, et al. (1989) p. 246
**Not included in Lysaght, et al. (1989)*

Chow, S.L and Murdock, B.B. The effect of a subsidiary task on iconic memory. Memory and Cognition. 3, 678-688; 1975.

Dewar, R.E., Ells, J.E., and Mundy, G. Reaction time as an index of traffic sign perception. Human Factors. 18, 381-392; 1976.

Dornic, S. Language dominance, spare capacity and perceived effort in bilinguals. Ergonomics. 23, 369-377; 1980.

Figarola, T.R. and Billings, C.E. Effects of meprobamate and hypoxia on psychomotor performance. Aerospace Medicine. 37, 951-954; 1966.

Fleishman, E.A. The prediction of total task performance from prior practice on task components. Human Factors. 7, 18-27; 1965.

Gabriel, R.F. and Burrows, A.A. Improving time-sharing performance of pilots through training. Human Factors. 10, 33-40; 1968.

Goldstein, I.L. and Dorfman, P.W. Speed and load stress as determinants of performance in a time sharing task. Human Factors. 20, 603-609; 1978.

Heimstra, N.W. The effects of "stress fatigue" on performance in a simulated driving situation. Ergonomics. 13, 209-218; 1970.

Herman, L.M. Study of the single channel hypothesis and input regulation within a continuous, simultaneous task situation. Quarterly Journal of Experimental Psychology. 17, 37-46; 1965.

Hilgendorf, E.L. Information processing practice and spare capacity. Australian Journal of Psychology. 19, 241-251; 1967.

Hoffman, E.R. and Jorbert, P.N. The effect of changes in some vehicle handling variables on driver steering performance. Human Factors. 8, 245-263; 1966.

Hohmuth, A.V. Vigilance performance in a bimodal task. Journal of Applied Psychology. 54, 520-525; 1970.

Huddleston, J.H.F. and Wilson, R.V. An evaluation of the usefulness of four secondary tasks in assessing the effect of a lag in simulated aircraft dynamics. Ergonomics. 14, 371-380; 1971.

Kelley, C.R. and Wargo, M.J. Cross-adaptive operator loading tasks. Human Factors. 9, 395-404; 1967.

Knowles, W.B. Operator loading tasks. Human Factors, 5, 151-161; 1963.

Kramer, A.F., Wickens, C.D., and Donchin, E. Performance enhancements under dual-task conditions. Proceedings of the 20th Annual Conference on Manual Control (pp. 21-35). Moffett Field, CA: Ames Research Center; 1984.

Kyriakides, K. and Leventhal, H.G. Some effects of intrasound on task performance. Journal of Sound and Vibration. 50, 369-388; 1977.

Lindsay, P.H. and Norman, D.A. Short-term retention during a simultaneous detection task. Perception and Psychophysics. 5, 201-205; 1969.

Long, J. Effect on task difficulty on the division of attention between nonverbal signals: Independence or interaction? Quarterly Journal of Experimental Psychology. 28, 179-193; 1976.

Lysaght, R.J., Hill, S.G., Dick, A.O., Plamondon, B.D., Linton, P.M., Wierwille, W.W., Zaklad, A.L., Bittner, A.C., and Wherry, R.J. Operator workload: comprehensive review and evaluation of operator workload methodologies (Technical Report 851). Alexandria, VA: Army Research Institute for the Behavioral and Social Sciences; June 1989.

Malmstrom, F.V., Reed, L.E., and Randle, R.J. Restriction of pursuit eye movement range during a concurrent auditory task. Journal of Applied Psychology. 68, 565-571; 1983.

McGrath, J.J. Performance sharing in an audio-visual vigilance task. Human Factors. 7, 141-153; 1965.

Mitsuda, M. Effects of a subsidiary task on backward recall. Journal of Verbal Learning and Verbal Behavior. 7, 722-725; 1968.

Monty, R.A. and Ruby, W.J. Effects of added workload on compensatory tracking for maximum terrain following. Human Factors. 7, 207-214; 1965.

Putz, V.R. and Rothe, R. Peripheral signal detection and concurrent compensatory tracking. Journal of Motor Behavior. 6, 155-163; 1974.

Schori, T.R. and Jones, B.W. Smoking and workload. Journal of Motor Behavior. 7, 113-120; 1975.

Soliday, S.M. and Schohan, B. Task loading of pilots in simulated low-altitude high-speed flight. Human Factors. 7, 45-53; 1965.

Stager, P. and Muter, P. Instructions and information processing in a complex task. Journal of Experimental Psychology. 87, 291-294; 1971.

Tyler, D.M. and Halcomb, C.G. Monitoring performance with a time-shared encoding task. Perceptual and Motor Skills. 38, 383-386; 1974.

Wetherell, A. The efficacy of some auditory vocal subsidiary tasks as measures of the mental load on male and female drivers. Ergonomics. 24, 197-214; 1981.

Wright, P., Holloway, C.M. and Aldrich, A.R. Attending to visual or auditory verbal information while performing other concurrent tasks. Quarterly Journal of Experimental Psychology. 26, 454-463; 1974.

3.1.9.14 Multiple Task Performance Battery of Secondary Tasks

General description - The Multiple Task Performance Battery (MTPB) requires subjects to time-share three or more of the following tasks: (1) light and dial monitoring, (2) mental math, (3) pattern discrimination, (4) target identification, (5) group problem solving, and (6) two-dimensional compensatory tracking. The monitoring tasks are used as secondary tasks and performance associated with these tasks as measures of workload.

Strengths and limitations - Increasing the number of tasks being time-shared does increase the detection time associated with the monitoring task. The MTPB may be difficult to implement in nonlaboratory settings. Lysaght, et al. (1989) reported the results of one study (Alluisi, 1971) in which the MTPB was paired with itself. In dual-task performance, performance of both the primary and the secondary MTPB tasks degraded.

Data requirements - The MTPB requires individual programming and analysis of six tasks as well as coordination among them during the experiment.

Thresholds - Not stated.

Sources -

Alluisi, E.A. and Morgan, B.B. Effects on sustained performance of time-sharing a three-phase code transformation task (3P-Cotran). Perceptual and Motor Skills. 33, 639-651; 1971.

Lysaght, R.J., Hill, S.G., Dick, A.O., Plamondon, B.D., Linton, P.M., Wierwille, W.W., Zaklad, A.L., Bittner, A.C., and Wherry, R.J. Operator workload: comprehensive review and evaluation of operator workload methodologies (Technical Report 851).

Alexandria, VA: Army Research Institute for the Behavioral and Social Sciences; June 1989.

3.1.9.15 Occlusion Secondary Task

General description - "The subject's view of a visual display is obstructed (usually by a visor). These obstructions are either initiated by the subject or imposed by the experimenter in order to determine the viewing time needed to perform a task adequately" (Lysaght, et al., 1989, p. 234).

Strengths and limitations - This task can be extremely disruptive of primary-task performance.

Based on the results of four studies in which a secondary occlusion task was used, Lysaght, et al. (1989) reported that performance remained stable on monitoring and driving primary tasks; and degraded on driving primary tasks. Performance of the secondary occlusion task degraded when paired with primary driving tasks (see Table 11).

TABLE 11.
References Listed by the Effect on Performance of Primary Tasks Paired with a Secondary Occlusion Task

	PRIMARY TASK			SECONDARY TASK		
Type	Stable	Degraded	Enhanced	Stable	Degraded	Enhanced
Driving	Farber and Gallagher (1972)	Hicks and Wierwille (1979) Senders, Kristofferson, Levison, Dietrich, and Ward (1967)			Farber and Gallagher (1972) Senders, *et al.* (1967)	
Monitoring	Gould and Schaffer (1967)					

from Lysaght, et al. (1989) p. 250

Data requirements - The following data are used to assess performance of this task: mean voluntary occlusion time and percent looking time/total time (Lysaght, et al., 1989, p. 236).

Thresholds - Not stated.

Sources -

Farber, E. and Gallagher, V. Attentional demand as a measure of the influence of visibility conditions on driving task difficulty. Highway Research Record. 414, 1-5; 1972.

Gould, J.D. and Schaffer, A. The effects of divided attention on visual monitoring of multi-channel displays. Human Factors. 9, 191-202; 1967.

Hicks, T.G. and Wierwille, W.W. Comparison of five mental workload assessment procedures in a moving-base driving simulator. Human Factors. 21, 129-143; 1979.

Lysaght, R.J., Hill, S.G., Dick, A.O., Plamondon, B.D., Linton, P.M., Wierwille, W.W., Zaklad, A.L., Bittner, A.C., and Wherry, R.J. Operator workload: comprehensive review and evaluation of operator workload methodologies (Technical Report 851). Alexandria, VA: Army Research Institute for the Behavioral and Social Sciences; June 1989.

Senders, J.W., Kristofferson, A.B., Levison, W.H., Dietrich, C.W., and Ward, J.L. The attentional demand of automobile driving. Highway Research Record. 195, 15-33; 1967.

3.1.9.16 Problem-Solving Secondary Task

General description - "The subject engages in a task which requires verbal or spatial reasoning. For example, the subject might attempt to solve anagram or logic problems" (Lysaght, et al., 1989, p. 233).

Strengths and limitations - "This class of tasks is thought to impose heavy demands on central processing resources" (Lysaght, et al., 1989, p. 233). Based on eight studies in which a problem-solving secondary task was used, Lysaght, et al. (1989) reported performance remained stable on a primary monitoring task and degraded on driving, tracking, and memory primary tasks. Performance of the secondary problem-solving task remained stable when paired with a primary tracking task; degraded when paired with problem-solving, driving, choice RT, and memory primary tasks; and improved when paired with a primary monitoring task (see Table 12).

TABLE 12.
References Listed by the Effect on Performance of Primary Tasks Paired with a Secondary Problem-Solving Task

Type	PRIMARY TASK			SECONDARY TASK		
	Stable	Degraded	Enhanced	Stable	Degraded	Enhanced
Choice RT					Schouten, Kalsbeek, and Leopold (1962)	
Driving		Wetherell (1981)			Wetherell (1981)	
Memory		Trumbo, Noble, and Swink (1967)			Trumbo, *et al.* (1967)	
Monitoring	Gould and Schaffer (1967) Smith, Lucaccini, Groth, and Lyman (1966)					Smith, *et al.* (1966)
Problem-Solving					Chiles and Alluisi (1979)	
Tracking		Trumbo, *et al.* (1967)			Trumbo, *et al.* (1967)	

from Lysaght, et al. (1989) p. 250

Data requirements - The following data are used for these tasks: number of correct responses, number of incorrect responses, and mean RT for correct responses (Lysaght, et al., 1989, p. 233).

Thresholds - Not stated.

Sources -

Chiles, W.D. and Alluisi, E.A. On the specification of operator or occupational workload with performance-measurement methods. Human Factors. 21, 515-528; 1979.

Gould, J.D. and Schaffer, A. The effects of divided attention on visual monitoring of multichannel displays. Human Factors. 9, 191-202; 1967.

Lysaght, R.J., Hill, S.G., Dick, A.O., Plamondon, B.D., Linton, P.M., Wierwille, W.W., Zaklad, A.L., Bittner, A.C., and Wherry, R.J. Operator workload: comprehensive review and evaluation of operator workload methodologies (Technical Report 851). Alexandria, VA: Army Research Institute for the Behavioral and Social Sciences; June 1989.

Schouten, J.F., Kalsbeek, J.W.H., and Leopold, F.F. On the evaluation of perceptual and mental load. Ergonomics. 5, 251-260; 1962.

Smith, R.L., Lucaccini, L.F., Groth, H., and Lyman, J. Effects of anticipatory alerting signals and a compatible secondary task on vigilance performance. Journal of Applied Psychology. 50, 240-246; 1966.
Trumbo, D., Noble, M. and Swink, J. Secondary task interference in the performance of tracking tasks. Journal of Experimental Psychology. 73, 232-240; 1967.
Wetherell, A. The efficacy of some auditory-vocal subsidiary tasks as measures of mental load on male and female drivers. Ergonomics. 24, 197-214; 1981.

3.1.9.17 Production/Handwriting Secondary Task

General description - "The subject is required to produce spontaneous handwritten passages of prose" (Lysaght, et al., 1989, p. 234).

Strengths and limitations - "With primary tasks that impose a high workload, subject's handwriting is thought to deteriorate (i.e., semantic and grammatical errors) under such conditions" (Lysaght, et al., 1989, p. 234). Lysaght, et al. (1989) cite a study reported by Schouten, Kalsbeek, and Leopold (1962) in which a spontaneous writing secondary task was paired with a choice RT primary task. Performance on the secondary task degraded.

Data requirements - The number of semantic and grammatical errors is used as data for this task (Lysaght, et al., 1989, p. 236).

Thresholds - Not stated.

Sources -

Lysaght, R.J., Hill, S.G., Dick, A.O., Plamondon, B.D., Linton, P.M., Wierwille, W.W., Zaklad, A.L., Bittner, A.C., and Wherry, R.J. Operator workload: comprehensive review and evaluation of operator workload methodologies (Technical Report 851). Alexandria, VA: Army Research Institute for the Behavioral and Social Sciences; June 1989.
Schouten, J.F., Kalsbeek, J.W.H., and Leopold, F.F. On the evaluation of perceptual and mental load. Ergonomics. 15, 251-260; 1962.

3.1.9.18 Psychomotor Secondary Task

General description - "The subject must perform a psychomotor task such as sorting different types of metal screws by size" (Lysaght, et al., 1989, p. 233).

Strengths and limitations - "Tasks of this nature are thought to reflect psychomotor skills" (Lysaght, et al., 1989, p. 233). Based on three studies in which a psychomotor secondary task was used, Lysaght, et al. (1989) reported that performance of a tracking primary task degraded. Performance of the secondary psychomotor task degraded when paired with either a tracking or choice RT primary task (see Table 13).

TABLE 13.
References Listed by the Effect on Performance of Primary Tasks Paired with a Secondary Psychomotor Task

	PRIMARY TASK			SECONDARY TASK		
Type	Stable	Degraded	Enhanced	Stable	Degraded	Enhanced
Choice RT					Schouten, Kalsbeek, and Leopold (1962)	
Tracking		Bergeron (1968) Wickens (1976)			Bergeron (1968)	

from Lysaght, et al. (1989) p. 251

Data requirements - The number of completed items is used to assess performance of this task.

Thresholds - Not stated.

Sources -

Bergeron, H.P. Pilot response in combined control tasks. Human Factors. 10, 277-282; 1968.

Lysaght, R.J., Hill, S.G., Dick, A.O., Plamondon, B.D., Linton, P.M., Wierwille, W.W., Zaklad, A.L., Bittner, A.C., and Wherry, R.J. Operator workload: comprehensive review and evaluation of operator workload methodologies (Technical Report 851). Alexandria, VA: Army Research Institute for the Behavioral and Social Sciences; June 1989.

Schouten, J.F., Kalsbeek, J.W.H., and Leopold, F.F. On the evaluation of perceptual and mental load. Ergonomics. 5, 251-260; 1962.

Wickens, C.D. The effects of divided attention on information processing in manual tracking. Journal of Experimental Psychology: Human Perception and Performance. 2, 1-12; 1976.

3.1.9.19 Randomization Secondary Task

General description - "The subject must generate a random sequence of numbers, for example. It is postulated that with increased workload levels subjects will generate repetitive responses (i.e., lack randomness in responses)" (Lysaght, et al., 1989, p. 232).

Strengths and limitations - The task is extremely intrusive and calculating "randomness" difficult and time consuming. Based on five studies that used a randomization secondary task, Lysaght, et al. (1989) reported performance remained stable on tracking, card sorting, and driving primary tasks; and degraded on tracking and memory primary tasks. Performance of the secondary randomization task remained stable when paired with a tracking primary task and degraded when paired with tracking and card-sorting primary tasks (see Table 14).

TABLE 14.
References Listed by the Effect on Performance of Primary Tasks Paired with a Secondary Randomization Task

	PRIMARY TASK			SECONDARY TASK		
Type	**Stable**	**Degraded**	**Enhanced**	**Stable**	**Degraded**	**Enhanced**
Card Sorting	Baddeley (1966)				Baddeley (1966)	
Driving	Wetherell (1981)					
Memory		Trumbo and Noble (1970)				
Tracking	Zeitlin and Finkelman (1975)	Truijens, Trumbo, and Wagenaar (1976)		Zeitlin and Finkelman (1975)	Truijens, *et al.* (1976)	

from Lysaght, et al. (1989) p. 250

Data requirements - The experimenter must calculate a percent redundancy score in bits of information.

Thresholds - Not stated.

Sources - Baddeley, A.D. The capacity for generating information by randomization. Quarterly Journal of Experimental Psychology. 18, 119-130; 1966.

Lysaght, R.J., Hill, S.G., Dick, A.O., Plamondon, B.D., Linton, P.M., Wierwille, W.W., Zaklad, A.L., Bittner, A.C., and Wherry, R.J. Operator workload: comprehensive review and evaluation of operator workload methodologies (Technical Report 851). Alexandria, VA: Army Research Institute for the Behavioral and Social Sciences; June 1989.

Truijens, C.L., Trumbo, D.A., and Wagenaar, W.A. Amphetamine and barbituate effects on two tasks performed singly and in combination. Acta Psychologica. 40, 233-244; 1976.

Trumbo, D. and Noble, M. Secondary task effects on serial verbal learning. Journal of Experimental Psychology. 85, 418-424; 1970.

Wetherell, A. The efficacy of some auditory-vocal subsidiary tasks as measures of the mental load on male and female drivers. Ergonomics. 24, 197-214; 1981.

Zeitlin, L.R. and Finkelman, J.M. Research note: Subsidiary task techniques of digit generation and digit recall indirect measures of operator loading. Human Factors. 17, 218-220; 1975.

3.1.9.20 Reading Secondary Task

General description - Subjects are asked to read digits or words aloud from a visual display. Measures can include: number of digits or words read, longest interval between spoken responses, longest string of consecutive digits or words, and the number of times three consecutive digits or words (Savage, Wierwille, and Cordes, 1978).

Strengths and limitations - This task has been sensitive to difficulty of a monitoring task. There were significant differences in the number of random digits spoken, the longest consecutive string of spoken digits, and the number of times three consecutive digits spoken (Savage, Wierwille, and Cordes, 1978). The longest interval between spoken responses was not significantly different among various levels of primary task difficulty (i.e., monitoring of two, three, or four meters).

Data requirements - Spoken responses must be recorded, timed, and tabulated.

Thresholds - Not stated.

Source -

Savage, R.E., Wierwille, W.W., and Cordes, R.E. Evaluating the sensitivity of various measures of operator workload using random digits as a secondary task. Human Factors. 20(6); 649-654; 1978.

3.1.9.21 Simple Reaction-Time Secondary Task

General description - "The subject is presented with one discrete stimulus (either visual or auditory) and generates one response to this stimulus" (Lysaght, et al., 1989, p. 232).

Strengths and limitations - This task minimizes central processing and response selection demands on the subject (Lysaght, et al., 1989, p. 232).

Based on ten studies in which a simple RT secondary task was used, Lysaght, et al. (1989) reported that performance remained stable on choice RT and classification primary tasks; degraded on tracking, classification, and lexical decision tasks; and improved on detection and driving primary tasks. Performance of the secondary simple RT task degraded when paired with tracking, choice RT, memory, detection, classification, driving, and lexical decision primary tasks; and improved when paired with a tracking primary task (see Table 15).

TABLE 15.
References Listed by the Effect on Performance of Primary Tasks Paired with a Secondary Simple RT Task

Type	PRIMARY TASK			SECONDARY TASK		
	Stable	Degraded	Enhanced	Stable	Degraded	Enhanced
Choice RT	Becker (1976)				Becker (1976)	
Classification	Comstock (1973)	Miller (1975)			Comstock (1973) Miller (1975)	
Detection			Laurell and Lisper (1978)		Laurell and Lisper (1978)	
Driving			Laurell and Lisper (1978)		Laurell and Lisper (1978) Lisper, Laurell, and Stening (1973)	
Lexical Decision		Becker (1976)			Becker (1976)	
Memory					Martin and Kelly (1974)	
Simulated Flight Task		Andre, Heers, and Cashion (1995)*				
Tracking		Heimstra (1970) Kelly and Klapp (1985) Klapp, Kelly, Battiste, and Dunbar (1984) Wickens and Gopher (1977)			Wickens and Gopher (1977)	Heimstra (1970)

from Lysaght, et al. (1989) p. 251
*Not included in Lysaght, et al. (1989)

Andre, Heers, and Cashion (1995) reported significant increase in pitch, roll, and yaw error of a simulated flight task while performing a secondary simple RT task.

Data requirements - The experimenter must be able to calculate mean RT for correct responses and the number of correct responses.

Thresholds - Not stated.

Sources -

Andre, A.D., Heers, S.T., and Cashion, P.A. Effects of workload preview on task scheduling during simulated instrument flight. International Journal of Aviation Psychology. 5(1), 5-23; 1995.

Becker, C.A. Allocation of attention during visual word recognition. Journal of Experimental Psychology: Human Perception and Performance. 2, 556-566; 1976.

Comstock, E.M. Processing capacity in a letter-matching task. Journal of Experimental Psychology. 100, 63-72; 1973.

Heimstra, N.W. The effects of “stress fatigue” on performance in a simulated driving situation. Ergonomics. 13, 209-213; 1970.

Kelly, P.A. and Klapp, S.T. Hesitation in tracking induced by a concurrent manual task. Proceedings of the 21st Annual Conference on Manual Control (pp. 19.1-19.3). Columbus, OH: Ohio State University; 1985.

Klapp, S.T., Kelly, P.A., Battiste, V., and Dunbar, S. Types of tracking errors induced by concurrent secondary manual task. Proceedings of the 20th Annual Conference on Manual Control (pp. 299-304). Moffett Field, CA: Ames Research Center; 1984.

Laurell, H. and Lisper, H.L. A validation of subsidiary reaction time against detection of roadside obstacles during prolonged driving. Ergonomics. 21, 81-88; 1978.

Lisper, H.L., Laurell, H., and Stening, G. Effects of experience of the driver on heart-rate, respiration-rate, and subsidiary reaction time in a three-hour continuous driving task. Ergonomics. 16, 501-506; 1973.

Lysaght, R.J., Hill, S.G., Dick, A.O., Plamondon, B.D., Linton, P.M., Wierwille, W.W., Zaklad, A.L., Bittner, A.C., and Wherry, R.J. Operator workload: comprehensive review and evaluation of operator workload methodologies (Technical Report 851). Alexandria, VA: Army Research Institute for the Behavioral and Social Sciences; June 1989.

Martin, D.W. and Kelly, R.T. Secondary task performance during directed forgetting. Journal of Experimental Psychology. 103, 1074-1079; 1974.

Miller, K. Processing capacity requirements of stimulus encoding. Acta Psychologica. 39, 393-410; 1975.

Wickens, C.D. and Gopher, D. Control theory measures of tracking as indices of attention allocation strategies. Human Factors, 19, 349-365; 1977.

3.1.9.22 Simulated Flight Secondary Task

General description - "Depending on the purpose of the particular study, the subject is required to perform various maneuvers (e.g., landing approaches) under different types of conditions such as instrument flight rules or simulated crosswind conditions" (Lysaght, et al., 1989, p. 234).

Strengths and limitations - This task requires extensive subject training.

Data requirements - The experimenter should record: mean error from required altitude, root-mean-square localizer error, root-mean-square glide-slope error, number of control movements, attitude high-pass mean square, and attitude high-pass mean square (Lysaght, et al., 1989, p. 236).

Thresholds - Not stated.

Sources -

Lysaght, R.J., Hill, S.G., Dick, A.O., Plamondon, B.D., Linton, P.M., Wierwille, W.W., Zaklad, A.L., Bittner, A.C., and Wherry, R.J. Operator workload: comprehensive review and evaluation of operator workload methodologies (Technical Report 851). Alexandria, VA: Army Research Institute for the Behavioral and Social Sciences; June 1989.

3.1.9.23 Spatial-Transformation Secondary Task

General description - "The subject must judge whether information (data)—provided by an instrument panel or radar screen—matches information which is spatially depicted by pictures or drawings of aircraft" (Lysaght, et al., 1989, p. 233). Lysaght, et al. (1989) cite work by Vidulich and Tsang (1985) in which performance of a spatial transformation secondary task was degraded when paired with a primary tracking task.

Kramer, Wickens, and Donchin (1984) reported a significant decrease in tracking error on a primary task when subjects performed translational changes of the cursor as a secondary task.

Strengths and limitations - "This task involves perceptual and cognitive processes" (Lysaght, et al., 1989, p. 233).

Data requirements - The following data are used to assess performance of this task: mean RT for correct responses, number of correct responses, and number of incorrect responses (Lysaght, et al., 1989, p. 236).

Thresholds - Not stated.

Sources -

Kramer, A.F., Wickens, C.D., and Donchin, E. Performance enhancements under dual-task conditions. Annual Conference on Manual Control, 1984, 21-35.

Lysaght, R.J., Hill, S.G., Dick, A.O., Plamondon, B.D., Linton, P.M., Wierwille, W.W., Zaklad, A.L., Bittner, A.C., and Wherry, R.J. Operator workload: comprehensive review and evaluation of operator workload methodologies (Technical Report 851). Alexandria, VA: Army Research Institute for the Behavioral and Social Sciences; June 1989.

Vidulich, M.A. and Tsang, P.S. Evaluation of two cognitive abilities tests in a dual-task environment. Proceedings of the 21st Annual Conference on Manual Control. 12.1-12.10; 1985.

3.1.9.24 Speed-Maintenance Secondary Task

General description - "The subject must operate a control knob to maintain a designated constant speed. This task is a psychomotor type task" (Lysaght, et al., 1989, p. 234).

Strengths and limitations - This task provides a constant estimate of reserve response capacity but may be extremely intrusive on primary-task performance.

Data requirements - Response is used as data for this task.

Thresholds - Not stated.

Source -

Lysaght, R.J., Hill, S.G., Dick, A.O., Plamondon, B.D., Linton, P.M., Wierwille, W.W., Zaklad, A.L., Bittner, A.C., and Wherry, R.J. Operator workload: comprehensive review and evaluation of operator workload methodologies (Technical Report 851). Alexandria, VA: Army Research Institute for the Behavioral and Social Sciences; June 1989.

3.1.9.25 Sternberg Memory Secondary Task

General description - The Sternberg (1966) recognition task presents a subject with a series of single letters. After each letter, the subject indicates whether that letter was or was not part of a previously memorized set of letters. RT is typically measured at two memory set sizes, two and four, and is plotted against set size (see Figure 3). Differences in the slope (b in Figure 3) across various design configurations indicate differences in central information processing demands. Changes in the intercept (a1 and a2 in Figure 3) suggest differences in either sensory or response demands. Additional data for this task include: number of correct responses and RTs for correct responses.

Strengths and limitations - The Sternberg task is sensitive to workload imposed by wind conditions, handling qualities, and display configurations. For example, Wolf (1978) reported the longest RTs to the Sternberg task occurred in the high gust, largest memory set (4), and poorest handling qualities condition. Schiflett (1980) reported both increased Sternberg RT and errors with degraded handling qualities during approach and landing tasks. Similarly, Schiflett, Linton, and Spicuzza (1982) reported increases in four Sternberg measures (RT for correct responses, intercept, slope, and percent errors) as handling qualities were degraded. Poston and Dunn (1986) used the Sternberg task to assess a kinesthetic tactile display. They recorded response speed and accuracy.

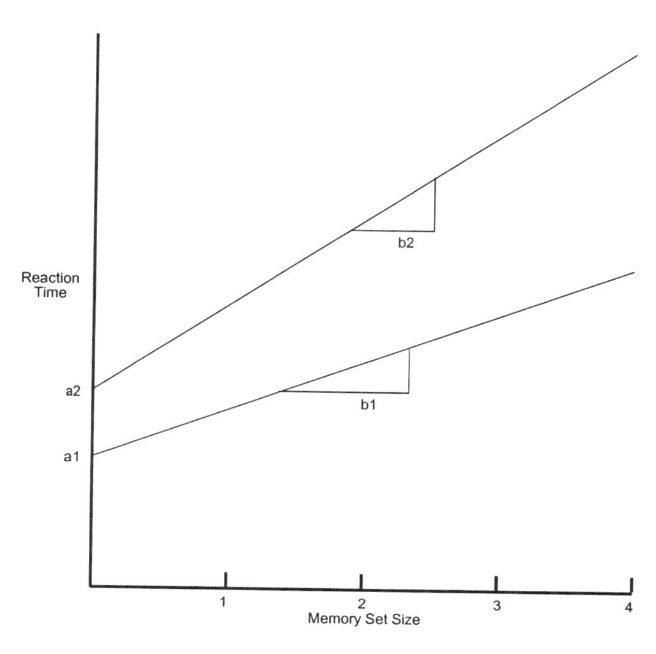

FIG. 3. Sternberg Memory Task Data

A second advantage of the task is that it is minimally intrusive on performance of a primary flight task (Schiflett, Linton, and Spicuzza, 1980; Dellinger and Taylor, 1985). Used alone the Sternberg provided reliable measure of mercury exposure (Smith and Langolf, 1981).

However, there are several problems associated with the use of the task. Based on data collected in a flight simulator (Taylor, Dellinger, Richardson, Weller, Porges, Wickens, LeGrand, and Davis, 1985), RTs increase with workload but more so for negative than positive responses. Further, work by Micalizzi (1981) suggests that performance of the Sternberg task is poorest when it is timeshared with a failure-detection task. Gawron, Schiflett, Miller, Ball, Slater, Parker, Lloyd, Travale, and Spicuzza (1988) used the Sternberg memory search task to access the inflight effects of pyridostigmine bromide. There were no significant effects of drug or crew position (pilot or copilot). This finding may be due to a true lack of a drug or crew position effect or to insensitivity of the measure. In an earlier study, Knotts and Gawron (1983) reported that the presence of the Peripheral Vision Display (PVD) reduced Sternberg RTs for one subject but not for another. They also reported that performance of the Sternberg task improved throughout the flight program and suggested extensive training before using the task in flight.

Micalizzi and Wickens (1980) compared Sternberg RT under single and dual task conditions. There were three manipulations of the Sternberg task: no mask, single mask, or double mask. They reported a significant increase in Sternberg RT under dual task condition for the two masked conditions as compared to the no mask condition.

Based on the results of four studies in which a secondary Sternberg task was used, Lysaght, et al. (1989) reported performance degraded on tracking, choice RT, and driving primary tasks; and improved on a tracking primary task. Performance of the secondary Sternberg task degraded when paired with a primary tracking task (see Table 16).

TABLE 16.
References Listed by the Effect on Performance of Primary Tasks Paired with a Secondary Task

Type	PRIMARY TASK			SECONDARY TASK		
	Stable	Degraded	Enhanced	Stable	Degraded	Enhanced
Choice RT		Hart, Shively, Vidulich, and Miller (1985)				
Driving		Wetherell (1981)				
Mental Mathematics		Payne, Peters, Birkmire, Bonto, Anatasi, and Wenger (1994)*				
Simple RT		Payne, Peters, Birkmire, Bonto, Anatasi, and Wenger (1994)*				
Simulated Flight	Crawford, Pearson, and Hoffman (1978)* Wierwille and Connor (1983)	O'Donnell (1976)*			Schiflett, Linton, and Spicuzza (1982)*	
Tracking	Tsang, Velaquez, and Vidulich (1996)	Wickens and Yeh (1985) Vidulich and Wickens (1986)	Briggs, Peters, and Fisher (1972)		Briggs, *et al.* (1972) Wickens and Yeh (1985)	

from Lysaght, et al. (1989) p. 252
**Not included in Lysaght, et al. (1989)*

Vidulich and Wickens (1986) reported that the "addition of the tracking task tends to overwhelm distinctions among the Sternberg task configurations" (p. 293). Further, Manzey, Lorenz, Schiewe, Finell, and Thiele (1995) reported significant differences in dual-task performance of the Sternberg and tracking prior and during space flight. However, there was no difference in single-task performance of the Sternberg.

Data requirements - The slopes and intercepts of the RT are used as data for this task. Wickens, Hyman, Dellinger, Taylor, and Meador (1986) have recommended that the Sternberg letters in the memorized set be changed after 50 trails. Schiflett (1983) recommended use of an adaptive interstimulus interval (ISI). Adaptation would be based on response accuracy. Knotts and Gawron (1983) recommended extensive training on the Sternberg task before data collection.

Thresholds - Not stated.

Sources -

Briggs, G.E., Peters, G.L., and Fisher, R.P. (1972). On the locus of the divided-attention effects. Perception and Psychophysics. 11, 315-320; 1972.

Crawford, B.M., Pearson, W.H., and Hoffman, M. Multipurpose digital switching and flight control workload (AMRL-TR-78-43). Wright-Patterson Air Force Base, OH: Air Force Aerospace Medical Research Laboratory, 1978.

Dellinger, J.A. and Taylor, H.L. The effects of atropine sulfate on flight simulator performance on respiratory period interactions. Aviation, Space, and Environmental Medicine; 1985.

Gawron, V.J., Schiflett, S., Miller, J., Ball, J., Slater, T., Parker, F., Lloyd, M., Travale, D., and Spicuzza, R.J. The effect of pyridostigmine bromide on inflight aircrew performance (USAFSAM-TR-87-24). Brooks AFB, TX: School of Aerospace Medicine; January 1988.

Hart, S.G., Shively, R.J., Vidulich, M.A., and Miller, R.C. The effects of stimulus modality and task integrity: Predicting dual-task performance and workload from single-task levels. Proceedings of the 21st Annual Conference on Manual Control (pp. 5.1-5.18). Columbus, OH: NASA Ames Research Center and Ohio State University; 1985.

Knotts, L.H. and Gawron, V.J. A preliminary flight evaluation of the peripheral vision display using the NT-33A aircraft (Report 6645-F-13). Buffalo, NY: Calspan; December 1983.

Lysaght, R.J., Hill, S.G., Dick, A.O., Plamondon, B.D., Linton, P.M., Wierwille, W.W., Zaklad, A.L., Bittner, A.C., and Wherry, R.J. Operator workload: comprehensive review and evaluation of operator workload methodologies (Technical Report 851). Alexandria, VA: Army Research Institute for the Behavioral and Social Sciences; June 1989.

Manzey, D., Lorenz, B., Schiewe, A., Finell, G., and Thiele, G. Dual-task performance in space: results from a single-case study during a short-term space mission. Human Factors. 37(4); 667-681, 1995.

Micalizzi, J. The structure of processing resource demands in monitoring automatic systems (Technical Report 81-2T). Wright-Patterson AFB, OH: Air Force Institute of Technology; 1981.

Micalizzi, J. and Wickens, C.D. The application of additive factors methodology to workload assessment in a dynamic system monitoring task. Champaign, IL: University of Illinois, Engineering-Psychology Research Laboratory, TREPL-80-2/ONR- 80-2, December 1980.

O'Donnell, R.D. Secondary task assessment of cognitive workload in alternative cockpit configurations. In B.O. Hartman (Ed.) Higher mental functioning in operational environments (pp. C10/1-C10/5). AGARD Conference Proceedings Number 181. Neuilly sur Seine, France: Advisory Group for Aerospace Research and Development; 1976.

Payne, D.G., Peters, L.J., Birkmire, D.P., Bonto, M.A., Anatasi, J.S., and Wenger, M.J. Effects of speech intelligibility level on concurrent visual task performance. Human Factors. 36(3); 441-475, 1994.

Poston, A.M. and Dunn, R.S. Helicopter flight evaluation of kinesthetic tactual displays: an interim report (HEL-TN-3-86). Aberdeen Proving Ground, MD: Human Engineering Laboratory; March 1986.

Schiflett, S.G. Evaluation of a pilot workload assessment device to test alternate display formats and control handling qualities. Patuxent River, MD: Naval Air Test Center, SY-33R-80, July 1980.

Schiflett, S.G. Theoretical development of an adaptive secondary task to measure pilot workload for flight evaluations. Proceedings of the 27th Annual Meeting of the Human Factors Society. 602-607; 1983.

Schiflett, S., Linton, P.M, and Spicuzza, R.J Evaluation of a pilot workload assessment device to test alternate display formats and control handling qualities. Proceedings of North Atlantic Treaty Organization (NATO) Advisory Group for Aerospace Research and Development (AGARD) (Paper Number 312). Neuilly-sur-Seine, France: AGARD; 1980.

Schiflett, S.G., Linton, P.M., and Spicuzza, R.J. Evaluation of a pilot workload assessment device to test alternative display formats and control handling qualities. Proceedings of the AIAA Workshops on flight testing to identify pilot workload and pilot dynamics (pp. 222-233). Edwards Air Force Base, CA: Air Force Flight Test Center; 1982.

Smith, P.J. and Langolf, G.D. The use of Sternberg's memory-scanning paradigm in assessing effects of chemical exposure. Human Factors. 23(6), 701-708; 1981.

Sternberg, S. High speed scanning in human memory. Science. 153, 852-654; 1966.

Taylor, H.L., Dellinger, J.A., Richardson, B.C., Weller, M.H., Porges, S.W., Wickens, C.D., LeGrand, J.E., and Davis, J.M. The effect of atropine sulfate on aviator performance (Technical Report APL-TR-85-1). Champaign, IL: University of Illinois, Aviation Research Laboratory; March 1985.

Tsang, P.S., Velaquez, V.L., and Vidulich, M.A. Viability of resource theories in explaining time-sharing performance. Acta Psychologica. 91(2), 175-206; 1996.

Vidulich, M.A. and Wickens, C.D. Causes of dissociation between subjective workload measures and performance. Applied Ergonomics. 17(4), 291-296; 1986.

Wetherell, A. The efficacy of some auditory-vocal subsidiary tasks as measures of the mental load on male and female drivers. Ergonomics. 24, 197-214; 1981.

Wickens, C.D., Hyman, F., Dellinger, J., Taylor, H., and Meador, M. The Sternberg memory search task as an index of pilot workload. Ergonomics. 29, 1371-1383; 1986.

Wickens, C.D. and Yeh, Y. POCs and performance decrements: A reply to Kantowitz and Weldon. Human Factors. 27, 549-554; 1985.

Wierwille, W.W. and Connor, S. Evaluation of 20 workload measures using a psychomotor task in a moving base aircraft simulator. Human Factors. 25, 1-16; 1983.

Wolf, J.D. Crew workload assessment: Development of a measure of operator workload (AFFDL-TR-78-165). Wright-Patterson Air Force Base, OH: Air Force Flight Dynamics Laboratory; December 1978.

3.1.9.26 Three-Phase Code Transformation Secondary Task

General description - "The subject operates the 3P-Cotran which is a workstation consisting of three indicator lights, a response board for subject responses and a memory unit that the subject uses to save his/her responses. The subject must engage in a 3 phase problem solving task by utilizing information provided by the indicator lights and recording solutions onto the memory unit" (Lysaght, et al., 1989, p. 234).

Strengths and limitations - "It is a synthetic work battery used to study work behavior and sustained attention" (Lysaght, et al., 1989, p. 234).

Data requirements - The following data are used to evaluate performance of this task: mean RT for different phases of response required and number of errors (resets) for different phases of response required (Lysaght, et al., 1989, p. 236).

Thresholds - Not stated.

Source -

Lysaght, R.J., Hill, S.G., Dick, A.O., Plamondon, B.D., Linton, P.M., Wierwille, W.W., Zaklad, A.L., Bittner, A.C., and Wherry, R.J. Operator workload: comprehensive review and evaluation of operator workload methodologies (Technical Report 851). Alexandria, VA: Army Research Institute for the Behavioral and Social Sciences; June 1989.

3.1.9.27 Time-Estimation Secondary Task

General description - Subjects are asked to produce a given time interval, usually 10 seconds, from the start of a stimulus, usually a tone. Measures for this task include the number of incomplete estimates and/or the length of the estimates.

Strengths and limitations - The technique is sensitive to workload (Hart, 1978; Wierwille, Casali, Connor, and Rahimi, 1985). For example, Bortolussi, Kantowitz, and Hart (1986) found a significant increase in 10-second time production intervals between easy and difficult flight scenarios. Bortolussi, Hart, and Shively (1987) reported similar results for a 5-second time production task.

In addition, the length of the time interval produced decreased from the beginning to the end of each flight. Gunning (1978) asked pilots to indicate when ten seconds had passed after an auditory tone. He reported that both the number of incomplete estimates and the length of the estimates increased as workload increased. Similarly, Madero, Sexton, Gunning, and Moss (1979) reported that the number of incomplete time estimates increased over time in an aerial delivery mission. These researchers also calculated a time-estimation ratio (the length of the time estimate in flight divided by the baseline estimate). This measure was also sensitive to workload, with significant increases occurring between cruise and Initial Point (IP) and cruise and Computed Air Release Point (CARP).

Connor and Wierwille (1983) recorded time estimation mean, standard deviation, absolute error, and root-mean-square error of completed estimates during three levels of gust and aircraft stability (load). There was only one significant load effect: the standard deviation of time estimates decreased from the low to the medium load then increased from the medium to high load. This same measure was sensitive to communication load, danger, and navigation load. Specifically, Casali and Wierwille (1983) reported significant increases in time-estimation standard deviation as communication load increased. Casali and Wierwille (1984) found significant increases between low and high danger conditions. Wierwille, Rahimi, and Casali (1985) reported the same results for navigation load, Hartzell (1979) for difficulty of precision hover maneuvers in a helicopter simulator.

Hauser, Childress, and Hart (1983) reported less variability in time estimates made using a counting technique than those made without a counting technique. Bobko, Bobko, and Davis (1986) reported, however, that verbal time estimates of a fixed time interval decrease as screen size (0.13, 0.28, and 0.58 diagonal meters) increased. In addition, men gave significantly shorter time estimates than women did.

Based on the results of four studies in which a time-estimation secondary task was used, Lysaght, et al. (1989) reported performance on a primary flight simulation task remained stable but degraded on a primary monitoring task. Performance of the secondary time-estimation task degraded when paired with either a monitoring or a flight simulation primary task (see Table 17).

Data requirements - Although some researchers have reported significant differences in several time-estimation measures, the consistency of the findings using time-estimation standard deviation suggest that this may be the best time-estimation measure.

Thresholds - Not stated.

TABLE 17.
References Listed by the Effect on Performance of Primary Tasks Paired with a Secondary Task

Type	PRIMARY TASK			SECONDARY TASK		
	Stable	Degraded	Enhanced	Stable	Degraded	Enhanced
Flight Simulation	Bortolussi, Hart, and Shively (1987) Bortolussi, Kantowitz, and Hart (1986) Casali and Wierwille (1983)* Kantowitz, Bartolussi i, and Hart (1987)* Wierwille and Connor (1983)* Wierwille, Rahimi, and Casali (1985)				Bortolussi, *et al.* (1987, 1986) Gunning (1978)* Wierwille, *et al.* (1985)	
Monitoring		Liu and Wickens (1987)			Liu and Wickens (1987)	

from Lysaght, et al. (1989) p. 252
**Not included in Lysaght, et al. (1989)*

Sources -

Bobko, D.J., Bobko, P., and Davis, M.A. Effect of visual display scale on duration estimates. Human Factors. 28(2), 153-158; 1986.

Bortolussi, M.R., Hart, S.G., and Shively, R.J. Measuring moment-to-moment pilot workload using synchronous presentations of secondary tasks in a motion-base trainer. Proceedings of the Fourth Symposium on Aviation Psychology. Columbus, OH: Ohio State University; 1987.

Bortolussi, M.R., Kantowitz, B.H., and Hart, S.G. Measuring pilot workload in a motion base trainer: A comparison of four techniques. Applied Ergonomics. 17, 278-283; 1986.

Casali, J.G. and Wierwille, W.W. A comparison of rating scale, secondary task, physiological, and primary task workload estimation techniques in a simulated flight emphasizing communications load. Human Factors. 25, 623-641; 1983.

Casali, J.G. and Wierwille, W.W. On the comparison of pilot perceptual workload: A comparison of assessment techniques addressing sensitivity and intrusion issues. Ergonomics, 27, 1033-1050; 1984.

Connor, S.A. and Wierwille, W.W. Comparative evaluation of twenty pilot workload assessment measures using a psychomotor task in a moving base aircraft simulator (Report 166457). Moffett Field, CA: NASA Ames Research Center; January 1983.

Gunning, D. Time estimation as a technique to measure workload. Proceedings of the Human Factors Society 22nd Annual Meeting, 41-45; 1978.

Hart, S.G. Subjective time estimation as an index of workload. Proceedings of the symposium on man-system interface: advances in workload study (pp. 115-131); 1978.

Hartzell, E.J. Helicopter pilot performance and workload as a function of night vision symbologies. Proceedings of the 18th IEEE Conference on Decision and Control Volumes. 995-996; 1979.

Hauser, J.R., Childress, M.E., and Hart, S.G. Rating consistency and component salience in subjective workload estimation. Proceedings of the Annual Conference on Manual Control. 127-149, 1983.

AKantowitz, B.H., Bortolussi, M.R., and Hart, S.G. Measuring workload in a motion base simulation III. Synchronous secondary task. Proceedings of the Human Factors Society 31st Annual Meeting. Santa Monica, CA: Human Factors Society; 1987.
Liu, Y.Y. and Wickens, C.D. Mental workload and cognitive task automation: An evaluation of subjective and time estimation metrics (NASA 87-2). Campaign, IL: University of Illinois Aviation Research Laboratory; 1987.
Lysaght, R.J., Hill, S.G., Dick, A.O., Plamondon, B.D., Linton, P.M., Wierwille, W.W., Zaklad, A.L., Bittner, A.C., and Wherry, R.J. Operator workload: comprehensive review and evaluation of operator workload methodologies (Technical Report 851). Alexandria, VA: Army Research Institute for the Behavioral and Social Sciences; June 1989.
Madero, R.P., Sexton, G.A., Gunning, D., and Moss, R. Total aircrew workload study for the AMST (AFFDL-TR-79-3080, Volume 1). Wright-Patterson Air Force Base, OH: Air Force Flight Dynamics Laboratory; February 1979.
Wierwille, W.W., Casali, J.G., Connor, S.A., and Rahimi, M. Evaluation of the sensitivity and intrusion of mental workload estimation technique. In W. Roner (Ed.), Advances in man-machine systems research (Volume 2, pp. 51-127). Greenwich, CT: JAI Press; 1985.
Wierwille, W.W. and Connor, S.A. Evaluation of 20 workload measures using a psychomotor task in a moving base aircraft simulator. Human Factors. 25, 1-16; 1983.
Wierwille, W.W., Rahimi, M., and Casali, J.G. Evaluation of 16 measures of mental workload using a simulated flight task emphasizing mediational activity. Human Factors. 27, 489-502; 1985.

3.1.9.28 Tracking Secondary Task

General description - "The subject must follow or track a visual stimulus (target) which is either stationary or moving by means of positioning an error cursor on the stimulus using a continuous manual response device" (Lysaght, et al., 1989, p. 232). Tracking tasks require nullifying an error between a desired and an actual location. The Critical Tracking Task (CTT) has been widely used.

Strengths and limitations - Tracking tasks provide a continuous measure of workload. This is especially important in subjects with large variations in workload. Several investigators (Corkindale, Cumming, and Hammerton-Fraser, 1969; Spicuzza, Pinkus, and O'Donnell, 1974) have used a secondary tracking task to successfully evaluate workload in an aircraft simulator. Spicuzza, et al. (1974) concluded that a secondary tracking task was a sensitive measure of workload. Clement (1976) used a cross-coupled tracking task in a STOL simulator to evaluate horizontal situation displays. Park and Lee (1992) reported tracking task performance distinguished passing and failing groups of flight students. Manzey, Lorenz, Schiewe, Finell, and Thiele (1995) reported significant decrements in performance of tracking and tracking with the Sternberg tasks between pre- and space-flight.

The technique may be useful in ground-based simulators but inappropriate for use in flight. For example, Ramacci and Rota (1975) required flight students to perform a secondary tracking task during their initial flights. The researchers were unable to quantitatively evaluate the scores on this task due to artifacts of air turbulence and subject fatigue. Further, Williges and Wierwille (1979) state that hardware constraints make in-flight use of a secondary tracking task unfeasible and potentially unsafe.

In addition, Damos, Bittner, Kennedy, and Harbeson (1981) reported that dual performance of a critical tracking task improved over 15 testing sessions suggesting a long training time to assymptote. Robinson and Eberts (1987) reported degraded performance on tracking when paired with a speech warning rather than a pictorial warning.

Andre, Heers, and Cashion (1995) reported increased pitch, roll, and yaw rmse in a primary simulated flight task when time-shared with a secondary target acquisition task.

Korteling (1991; 1993) reported age differences in dual task performance of two one-dimensional compensatory tracking tasks.

Based on the results of twelve studies in which a secondary tracking task was used, Lysaght, et al. (1989) reported that performance remained stable on tracking, monitoring, and problem-solving primary tasks; degraded on tracking, choice RT, memory, simple RT, detection, and classification primary tasks; and improved on a tracking primary task. Performance of the secondary tracking task degraded when paired with tracking, choice RT, memory, monitoring, problem-solving, simple RT, detection, and classification primary tasks and improved when paired with a tracking primary task (see Table 18).

TABLE 18.
References Listed by the Effect on Performance of Primary Tasks Paired with a Secondary Tracking Task

	PRIMARY TASK			SECONDARY TASK		
Type	Stable	Degraded	Enhanced	Stable	Degraded	Enhanced
Choice RT		Looper (1976) Whitaker (1979)			Hansen (1982) Whitaker (1979)	
Classification		Wickens, Mountford, and Schreiner (1981)			Wickens, *et al.* (1981)	
Detection		Wickens, *et al.* (1981)			Wickens, *et al.* (1981) Robinson and Eberts (1987)*	
Memory		Johnston, Greenberg, Fisher, and Martin (1970)			Johnston, *et al.* (1970)	
Monitoring	Griffiths and Boyce (1971)				Griffiths and Boyce (1971)	
Problem Solving	Wright, Holloway, and Aldrich (1974)				Wright, *et al.* (1974)	
Simple RT		Schmidt, Kleinbeck, and Brockman (1984)			Schmidt, *et al.* (1984)	
Simulated Flight Task		Andre, Heers, and Cashion (1995)*				
Tracking	Mirchandani (1972)	Gawron (1982)* Hess and Teichgraber (1974) Wickens and Kessel (1980) Wickens, Mountford, and Schreiner (1981)	Tsang and Wickens (1984)		Gawron (1982)* Tsang and Wickens (1984) Wickens, *et al.* (1981)	Mirchandani (1972)

from Lysaght, et al. (1989) p. 251
**Not included in Lysaght, et al. (1989)*

Data requirements - The experimenter should calculate: integrated errors in mils (root-mean-square error), total time on target, total time of target, number of times of target, and number of target hits (Lysaght, et al., 1989, p. 235). A secondary tracking task is most appropriate in systems when a continuous measure of workload is required. It is recommended that known forcing functions be used rather than unknown, quasi-random disturbances.

Thresholds - Not stated.

Sources -

Andre, A.D., Heers, S.T., and Cashion, P.A. Effects of workload preview on task scheduling during simulated instrument flight. International Journal of Aviation Psychology. 5(1), 5-23, 1995.

Clement, W.F. Investigating the use of a moving map display and a horizontal situation indicator in a simulated powered-lift short-haul operation. Proceedings of the 12th Annual NASA-University Conference on Manual Control. 201-224; 1976.

Corkindale, K.G.G., Cumming, F.G., and Hammerton-Fraser, A.M. Physiological assessment of a pilot's stress during landing. Proceedings of the NATO Advisory Group for Aerospace Research and Development. 56; 1969.

Damos, D., Bittner, A.C., Kennedy, R.S., and Harbeson, M.M. Effects of extended practice on dual-task tracking performance, Human Factors. 23(5), 625-631; 1981.

Gawron, V.J. Performance effects of noise intensity, psychological set, and task type and complexity. Human Factors. 24(2), 225-243; 1982.

Griffiths, I.D. and Boyce, P.R. Performance and thermal comfort. Ergonomics. 14, 457-468; 1971.

Hansen, M.D. Keyboard design variables in dual-task. Proceedings of the 18th Annual Conference on Manual Control (pp. 320-326). Dayton, OH: Flight Dynamics Laboratory; 1982.

Hess R.A. and Teichgraber, W.M. Error quantization effects in compensatory tracking tasks. IEEE Transactions on Systems, Man, and Cybernetics. SMC-4, 343-349; 1974.

Johnston, W.A., Greenberg, S.N., Fisher, R.P., and Martin, D.W. Divided attention: A vehicle for monitoring memory processes. Journal of Experimental Psychology. 83, 164-171; 1970.

Korteling, J.E. Effects of skill integration and perceptual competition on age-related differences in dual-task performance. Human Factors. 33(1), 35-44, 1991.

Korteling, J.E. Effects of age and task similarity on dual-task performance. Human Factors. 35(1); 1993.

Looper, M. The effect of attention loading on the inhibition of choice reaction time to visual motion by concurrent rotary motion. Perception and Psychophysics. 20, 80-84; 1976.

Lysaght, R.J., Hill, S.G., Dick, A.O., Plamondon, B.D., Linton, P.M., Wierwille, W.W., Zaklad, A.L., Bittner, A.C., and Wherry, R.J. Operator workload: comprehensive review and evaluation of operator workload methodologies (Technical Report 851). Alexandria, VA: Army Research Institute for the Behavioral and Social Sciences; June 1989.

Manzey, D., Lorenz, B., Schiewe, A., Finell, G., and Thiele, G. Dual-task performance in space: results from a single-case study during a short-term space mission. Human Factors. 37(4), 667-681; 1995.

Mirchandani, P.B. An auditory display in a dual axis tracking task. IEEE Transactions on Systems, Man, and Cybernetics. 2, 375-380; 1972.

Park, K.S. and Lee, S.W. A computer-aided aptitude test for predicting flight performance of trainees. Human Factors. 34(2), 189-204; 1992.

Ramacci, C.A. and Rota, P. Flight fitness and psycho-physiological behavior of applicant pilots in the first flight missions. Proceedings of NATO Advisory Group for Aerospace Research and Development. 153, B8; 1975.

Robinson, C.P. and Eberts, R.E. Comparison of speech and pictorial displays in a cockpit environment. Human Factors. 29(1): 31-44; 1987.

Schmidt, K.H., Kleinbeck, U., and Brockman, W. Motivational control of motor performance by goal setting in a dual-task situation. Psychological Research. 46, 129-141; 1984.

Spicuzza, R.J., Pinkus, A.R., and O'Donnell, R.D. Development of performance assessment methodology for the digital avionics information system. Dayton, OH: Systems Research Laboratories; August 1974.

Tsang, P.S. and Wickens, C.D. The effects of task structures on time-sharing efficiency and resource allocation optimality. Proceedings of the 20th Annual Conference on Manual Control (pp. 305-317). Moffett Field, CA: Ames Research Center; 1984.

Whitaker, L.A. Dual task interference as a function of cognitive processing load. Acta Psychologica. 43, 71-84; 1979.

Wickens, C.D. and Kessel, C. Processing resource demands of failure detection in dynamic systems. Journal of Experimental Psychology: Human Perception and Performance. 6, 564-577; 1980.

Wickens, C.D., Mountford, S.J., and Schreiner, W. Multiple resources, task-hemispheric integrity, and individual differences in time-sharing. Human Factors. 23, 211-229; 1981.

Williges, R.C. and Wierwille, W.W. Behavioral measures of aircrew mental workload. Human Factors. 21, 549-574; 1979.

Wright, P., Holloway, C.M., and Aldrich, A.R. Attending to visual or auditory verbal information while performing other concurrent tasks. Quarterly Journal of Experimental Psychology. 26, 454-463; 1974.

3.1.9.29 Workload Scale Secondary Task

General description - A workload scale was developed by tallying the number of persons who performed better in each task combination of the Multiple Task Performance Battery (MTPB) and converting these proportions to z scores (Chiles and Alluisi, 1979). The resulting z scores are multiplied by -1 so that the most negative score is associated with the lowest workload.

Strengths and limitations - Workload scales are easy to calculate but have two assumptions: (1) linear additivity and (2) no interaction between tasks. Some task combinations may violate these assumptions. Further, the intrusiveness of secondary tasks may preclude their use in nonlaboratory settings.

Data requirements - Performance of multiple combinations of tasks is required.

Thresholds - Dependent on task and task combinations being used.

Source -

Chiles, W.D. and Alluisi, E.A. On the specification of operator or occupational workload with performance-measurement methods. Human Factors. 21 (5), 515-528; 1979.

3.1.10 Task Difficulty Index

General description - The Task Difficulty Index was developed by Wickens and Yeh (1985) to categorize the workload associated with typical laboratory tasks. The index has four dimensions:

1) familiarity of stimuli
   0 = letters
   1 = spatial dot patterns, tracking cursor
2) number of concurrent tasks
   0 = single
   1 = dual
3) task difficulty
   0 = memory set size 2
   1 = set size 4, second-order tracking, delayed recall
4) resource competition
   0 = no competition
   1 = competition for either modality of stimulus (visual, auditory) or central processing (spatial, verbal)" (Gopher and Braune, 1984).

The Task Difficulty Index is the sum of the scores on each of the four dimensions listed above.

Strengths and limitations - Gopher and Braune (1984) reported a significant positive correlation (+0.93) between Task Difficulty Index and subjective measures of workload. Their data were based on responses from 55 male subjects performing 21 tasks including Sternberg, hidden pattern, card rotation tracking, maze tracing, delayed digit recall, and dichotic listening.

Data requirements - This method requires the user to describe the tasks to be performed on the four dimensions given above.

Thresholds - Values vary between 0 and 4.

Sources -

Gopher, D. and Braune, R. On the psychophysics of workload: Why bother with subjective measures? Human Factors. 26(5): 519-532; 1984.

Wickens, C.D. and Yeh, Y. POCs and performance decrments: A reply to Kantowitz and Weldon. Human Factors, 27, 549-554; 1985.

3.1.11 Time Margin

General description - After a review of current in-flight workload measures, Gawron, Schiflett, and Miller (1989) identified five major deficiencies: (1) the subjective ratings showed wide individual differences well beyond those that could be attributed to experience and ability differences; (2) most of the measures were not comprehensive and assessed only a single dimension of workload; (3) many workload measures were intrusive in terms of requiring task responses or subjective ratings or the use of electrodes; (4) some measures were confusing to subjects in high stress, for example, the meanings of ratings would be forgotten in high-workload environments, so lower than actual values would be given by the pilot; and (5) subjects would misperceive the number of tasks to be performed and provide an erroneously low measure of workload. Gawron then returned to the purpose of workload measure: to identify potentially dangerous situations. Poor designs, inadequate procedures, poor training, or the proximity to catastrophic conditions could induce such situations. The most objective measure of danger in a situation is time until the aircraft is destroyed if control action is not taken. These times include: time until impact, time until the aircraft is overstressed and breaks apart, and time until the fuel is depleted.

Strengths and limitations - The time-limit workload measure is quantitative, objective, directly related to performance, and can be tailored to any mission. For example, time until a surface-to-air missile destroys the aircraft is a good measure in air-to-ground penetration missions. In addition, the times can be easily computed from measures of aircraft performance. Finally, these times can be summed over intervals of any length to provide interval-by-interval workload comparisons.

Data requirements - This method is useful whenever aircraft performance data are available.

Thresholds - Minimum is 0, maximum is infinity.

Source -

Gawron, V.J., Schiflett, S.G., and , R.C. Measures of in-flight workload. In R.S. Jensen (Ed.) Aviation psychology. London: Gower; 1989.

3.2 Subjective Measures of Workload

Casali and Wierwille (1983) identified several advantages of subjective measures: "inexpensive, unobtrusive, easily administered, and readily transferable to full-scale aircraft and to a wide range of tasks" (p. 640). Gopher (1983) concluded that subjective measures "are well worth the bother" (p. 19). Wickens (1984) states that subjective measures have high face validity. Muckler and Seven (1992) state that subjective measures may be essential.

O'Donnell and Eggemeier (1986) however, identified six limitations of subjective measures of workload: (1) potential confound of mental and physical workload, (2) difficulty in distinguishing external demand/task difficulty from actual workload, (3) unconscious processing of information that the operator cannot rate subjectively, (4) dissociation of subjective ratings and task performance, (5) require well-defined question, and (6) dependence on short-term memory.

In addition, Meshkati, Hancock, and Rahimi (1990) warn that raters may interpret the words in a rating scale differently thus leading to inconsistent results.

Individual descriptions of subjective measures of workload are provided in the following sections. A summary is provided in Table 19.

Sources -

Casali, J.G. and Wierwille, W.W. A comparison of rating scale, secondary task, physiological, and primary-task workload estimation techniques in a simulated flight task emphasizing communications load. Human Factors. 25, 623-642; 1983.

Gopher, D. The workload book: assessment of operator workload to engineering systems (NASA-CR-166596). Moffett Field, CA: NASA Ames Research Center; November 1983.

Meshkati, N., Hancock, P.A., and Rahimi, M. Techniques in mental workload assessment. In J.R. Wilson and E.N. Corlett (Eds.). Evaluation of human work. A practical ergonomics methodology. New York: Taylor and Francis; 1990.

TABLE 19.
Comparison of Subjective Measures of Workload

Section	Measure	Reliability	Task Time	Ease of Scoring
3.2.1	Analytical Hierarchy Process	High	Requires rating pairs of tasks	Computer scored
3.2.2	Arbeitswissenshaftliches Erhebungsverfahren zur Tatigkeitsanalyze	High	Requires rating 216 items	Requires multivariate statistics
3.2.3	Bedford Workload Scale	High	Requires two decisions	No scoring needed
3.2.4	Computerized Rapid Analysis of Workload	Unknown	None	Requires detailed mission timeline
3.2.5	Continuous Subjective Assessment of Workload	High	Requires programming computer prompts	Computer scored
3.2.6	Cooper-Harper Rating Scale	High	Requires three decisions	No scoring needed
3.2.7	Crew Status Survey	High	Requires one decision	No scoring needed
3.2.8	Dynamic Workload Scale	High	Requires ratings by pilot and observer whenever workload changes	No scoring needed
3.2.9	Equal-Appearing Intervals	Unknown	Requires ratings in several categories	No scoring needed
3.2.10	Finegold Workload Rating Scale	High	Requires five ratings	Requires calculating an average
3.2.11	Flight Workload Questionnaire	May evoke response bias	Requires four ratings	No scoring needed
3.2.12	Hart and Bortolussi Rating Scale	Unknown	Requires one rating	No scoring needed
3.2.13	Hart and Hauser Rating Scale	Unknown	Requires six ratings	Requires interpolating quantity from mark on scale
3.2.14	Honeywell Cooper-Harper Rating Scale	Unknown	Requires three decisions	No scoring needed
3.2.15	Magnitude Estimation	Moderate	Requires comparison to a standard	No scoring needed
3.2.16	McCracken-Aldrich Technique	Unknown	May require months of preparation	Requires computer programmer
3.2.17	McDonnell Rating Scale	Unknown	Requires three or four decisions	No scoring needed
3.2.18	Mission Operability Assessment Technique	Unknown	Requires two ratings	Requires conjoint measurement techniques
3.2.19	Modified Cooper-Harper Rating Scale	High	Requires three decisions	No scoring needed
3.2.20	Multi-Descriptor Scale	Low	Requires six ratings	Requires calculating an average
3.2.21	Multidimensional Rating Scale	High	Requires eight ratings	Requires measurement of line length
3.2.22	NASA Bipolar Rating Scale	High	Requires ten ratings	Requires weighting procedure
3.2.23	NASA Task Load Index	High	Requires six ratings	Requires weighting procedure
3.2.24	Overall Workload Scale	Moderate	Requires one rating	No scoring needed
3.2.25	Pilot Objective/Subjective Workload Assessment Technique	High	Requires one rating	No scoring needed
3.2.26	Pilot Subjective Evaluation	Unknown	Requires rating systems on four scales and completion of questionnaire	Requires extensive interpretation
3.2.27	Profile of Mood States	High	Requires about ten minutes to complete	Requires manual or computer computer scoring
3.2.28	Sequential Judgment Scale	High	Requires rating each task	Requires measurement and conversion to percent
3.2.29	Subjective Workload Assessment Technique	High	Requires prior card sort and three ratings	Requires computer scoring
3.2.30	Subjective Workload Dominance Technique	High	Requires N(N-1)/2 paired comparisons	Requires calculating geometric means
3.2.31	Task Analysis Workload	Unknown	May require months of preparation	Requires detailed task analysis
3.2.32	Utilization	High	None	Requires regression
3.2.33	Workload/Compensation/Interference/Technical Effectiveness	Unknown	Requires ranking sixteen matrix cells	Requires complex mathematical processing
3.2.34	Zachery/Zaklad Cognitive Analysis	Unknown	May require months of preparation	Requires detailed task analysis

Muckler, F.A. and Seven, S.A. Selecting performance measures "objective" versus "subjective" measurement. Human Factors. 34:441-455; 1992.

O'Donnell, R.D. and Eggemeier, F.T. Workload assessment methodology. In K.R. Boff, L. Kaufman, and J.P. Thomas (Eds.) Handbook of perception and human performance. New York: Wiley and Sons; 1986.

Wickens, C.D. Engineering psychology and human performance. Columbus, OH: Charles E. Merrill; 1984.

3.2.1 Analytical Hierarchy Process

General description - The analytical hierarchy process (AHP) uses the method of paired comparisons to measure workload. Specifically, subjects rate which of a pair of conditions has the higher workload. All combinations of conditions must be compared. Therefore, if there are n conditions, the number of comparisons is 0.5n(n-1).

Strengths and limitations - Lidderdale (1987) found high consensus in the ratings of both pilots and navigators for a low-level tactical mission. Vidulich and Tsang (1987) concluded that AHP ratings were more valid and reliable than either an overall workload rating or NASA-TLX. Vidulich and Bortolussi (1988) reported that AHP ratings were more sensitive to attention than secondary RTs. Vidulich and Tsang (1988) reported high test/retest reliability. Bortolussi and Vidulich (1991) reported significantly higher workload using speech controls than manual controls in a combat helicopter simulated mission. AHP accounted for 64.2% the variance in mission phase (Vidulich and Bortolussi, 1988). AHP was also sensitive to degree of automation in a combat helicopter simulation (Bortolussi and Vidulich, 1989).

Metta (1993) used the AHP to develop a rank ordering of computer interfaces. She identified the following advantages of AHP: 1) easy to quantify consistency in human judgments, 2) yields useful results in spite of small sample sizes and low probability of statistical significant results, and 3) requires no statistical assumptions.

However, complex mathematical procedures must be employed (Lidderdale, 1987; Lidderdale and King, 1985; Saaty, 1980).

Data requirements - Four steps are required to use the AHP. First, a set of instructions must be written. A verbal review of the instructions should be conducted after the subjects have read the instructions to ensure their understanding of the task. Second, a set of evaluation sheets must be designed to collect the subjects' data. An example is presented in Figure 4. Each sheet has the two conditions to be compared in separate columns, one on the right side of the page, the other on the left. A 17-point rating scale is placed between the two sets of conditions. The scale uses five descriptors in a predefined order and allows a single point between each for mixed ratings. Vidulich (1988) defined the scale descriptors (see Table 20). Budescu, Zwick, and Rapoport (1986) provide critical value tables for detecting inconsistent judgments and subjects.

WORKLOAD JUDGMENTS

ABSOLUTE   VERY STRONG   STRONG   WEAK   EQUAL   WEAK   STRONG   VERY STRONG   ABSOLUTE

ILS APPROACH WITH HUD   _ _ _ _ _ _ _ _ _ _ _ _ _ _ _ _ _   ILS APPROACH WITHOUT HUD

FIG. 4. Example AHP Rating Scale

TABLE 20.
Definitions of AHP Scale Descriptors

EQUAL	The two task combinations are absolutely equal in the amount of workload generated by the simultaneous tasks.
WEAK	Experience and judgment slightly suggest that one of the combinations of tasks has more workload than the other.
STRONG	Experience and judgment strongly suggest that one of the combinations has higher workload.
VERY STRONG	One task combinations is strongly dominant in the amount of workload, and this dominance is clearly demonstrated in practice.
ABSOLUTE	The evidence supporting the workload dominance of one task combination is the highest possible order of affirmation (adapted from Vidulich, 1988, p. 5).

Third, the data must be scored. The scores range from +8 (absolute dominance of the left-side condition over the right-side condition) to -8 (absolute dominance of the right-side condition over the left-side condition). Finally, the scores are input, in matrix form, into a computer program. The output of this program is a scale weight for each condition and three measures of goodness of fit.

Thresholds - Not stated.

Sources -

Bortolussi, M.R. and Vidulich, M.A. The effects of speech controls on performance in advanced helicopters in a double stimulation paradigm. Proceedings of the International Symposium on Aviation Psychology. 216-221, 1991.

Bortolussi, M.R. and Vidulich, M.A. The benefits and costs of automation in advanced helicopters: An empirical study. Proceedings of the Fifth International Symposium on Aviation Psychology. 594-559, 1989.

Budescu, D.V., Zwick, R., and Rapoport, A. A comparison of the eigen value method and the geometric mean procedure for ratio scaling. Applied Psychological Measurement. 10, 68-78; 1986.

Lidderdale, I.G. Measurement of aircrew workload during low-level flight, practical assessment of pilot workload (AGARD-AG-282). Proceedings of NATO Advisory Group for Aerospace Research and Development (AGARD). Neuilly-sur-Seine, France: AGARD; 1987.

Lidderdale, I.G. and King, A.H. Analysis of subjective ratings using the analytical hierarchy process: A microcomputer program. High Wycombe, England: OR Branch NFR, HQ ST C, RAF; 1985.

Metta, D.A. An application of the Analytic Hierarchy Process: A rank-ordering of computer interfaces. Human Factors. 35(1), 141-157; 1993.

Saaty, T.L. The analytical hierarchy process. New York: McGraw-Hill; 1980.

Vidulich, M.A. Notes on the AHP procedure; 1988. (Available from Dr. Michael A. Vidulich, Human Engineering Division, AAMRL/HEG, Wright-Patterson Air Force Base, OH 45433-6573.)

Vidulich, M.A. and Bortolussi, M.R. A dissociation of objective and subjective workload measures in assessing the impact of speech controls in advanced helicopters. Proceedings of the Human Factors Society 32nd Annual Meeting. 1471-1475; 1988.

Vidulich, M.A. and Tsang, P.S. Absolute magnitude estimation and relative judgement approaches to subjective workload assessment. Proceedings of the Human Factors Society 31st Annual Meeting. 1057-1061; 1987.

Vidulich, M.A. and Tsang, P.S. Evaluating immediacy and redundancy in subjective workload techniques. Proceedings of the Twenty-Third Annual Conference on Manual Control; 1988.

3.2.2 Arbeitswissenshaftliches Erhebungsverfahren zur Tatigkeitsanalyze

General description - Arbeitswissenschaftliches Erhebungsverfahren zur Tatigkeitsanalyze (AET) was developed in Germany to measure workload. AET has three parts: (1) work system analysis, which rates the "type and properties of work objects, the equipment to be used as well as physical social and organizational work environment" (North and Klaus, 1980, p. 788) on both nominal and ordinal scales; (2) task analysis, which uses a 31-item ordinal scale to rate "material work objects, . abstract (immaterial) work objects, and man-related tasks" (p. 788); and (3) job-demand analysis, which is used to evaluate the conditions under which the job is performed. "The 216 items of the AET are rated on nominal or ordinal scales using five codes as indicated for each item: frequency, importance, duration, alternative and special (intensity) code" (p. 790).

Strengths and limitations - AET has been used in over 2,000 analyses of both manufacturing and management jobs.

Data requirements - Profile analysis is used to analyze the job workload. Cluster analysis is used to identify elements of jobs that "have a high degree of natural association among one another" (p. 790). Multivariate statistics are used for "placement, training, and job classification" (p. 790).

Thresholds - Not stated.

Source -

North, R.A. and Klaus, J. Ergonomics methodology - an obstacle or promoter for the implementation of ergonomics in industrial practice? Ergonomics. 23 (8); 781-795; 1980.

3.2.3 Bedford Workload Scale

General description - Roscoe (1984) described a modification of the Cooper-Harper scale created by trial and error with the help of test pilots at the Royal Aircraft Establishment at Bedford, England. The Bedford Workload Scale retained the binary decision tree and the four- and ten-rank ordinal structures of the Cooper-Harper scale (see Figure 5). The three-rank ordinal structure asked pilots to assess whether: (1) it was possible to complete the task, (2) the workload was tolerable, and (3) the workload was satisfactory without reduction. The rating-scale end points were: *workload insignificant* to *task abandoned.* In addition to the structure, the Cooper-Harper (1969) definition of pilot workload was used: "...the integrated mental and physical effort required to satisfy the perceived demands of a specified flight task" (Roscoe, 1984, p. 12-8). The concept of spare capacity was used to help define levels of workload.

Strengths and limitations - The Bedford Workload Scale was reported to be welcomed by pilots. Roscoe (1984) reported that pilots found the scale "easy to use without the need to always refer to the decision tree." He also noted that it was necessary to accept ratings of 3.5 from the pilots. These statements suggest that the pilots emphasized the ten- rather than the four-rank, ordinal structure of the Bedford Workload Scale. Roscoe (1984) found that pilot workload ratings and heart rates varied in similar manners

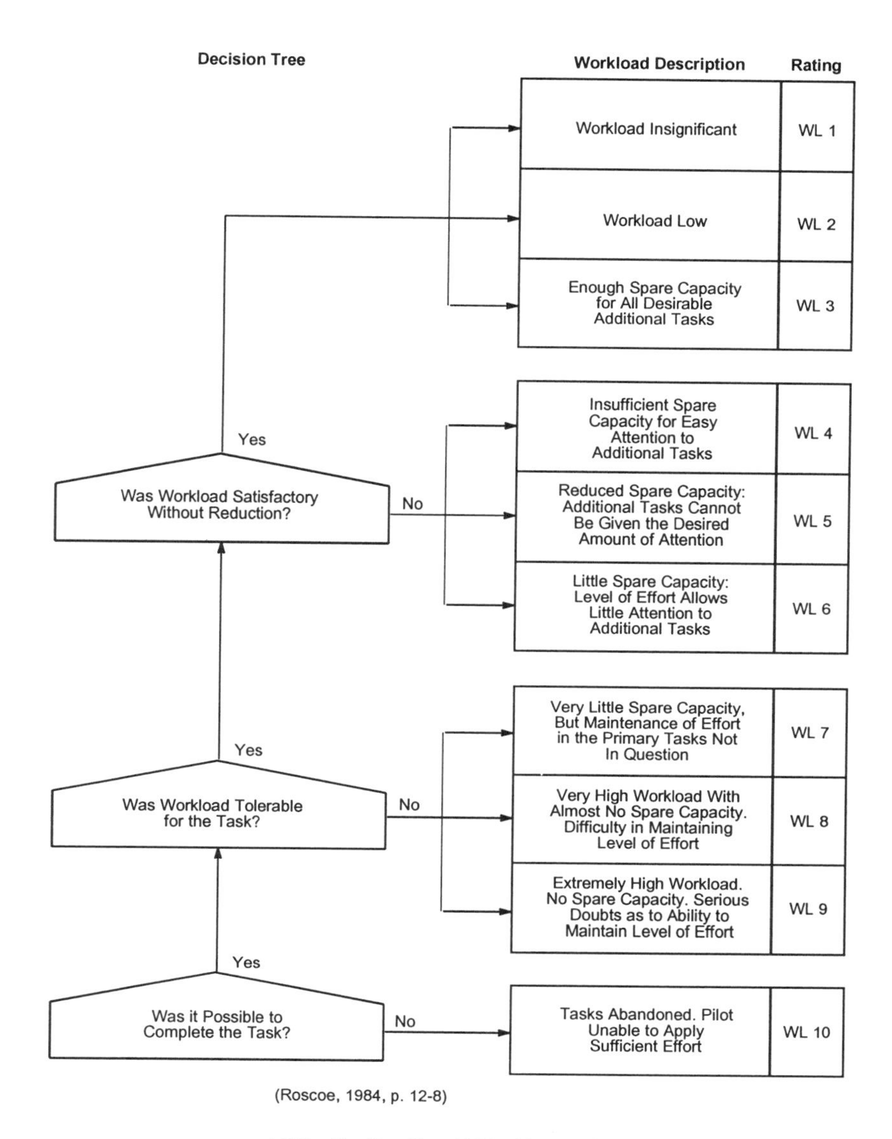

FIG. 5. Bedford Workload Scale

during close-coupled in-flight maneuvers in a BAe 125 twinjet aircraft. He felt that the heart-rate information complemented and increased the value of subjective workload ratings. He also noted the lack of absolute workload information provided by the Bedford Workload Scale and by heart-rate data. Wainwright (1987) used the scale during certification of the BAe 146 aircraft. Tsang and Johnson (1987) concluded that the Bedford Scale provided a good measure of spare capacity.

Roscoe (1987) reported that the scale was well accepted by aircrews. Corwin, Sandry-Garza, Biferno, Boucek, Logan, Jonsson, and Metalis (1989) concluded that the Bedford scale is a reliable and valid measure of workload based on flight simulator data. Vidulich and Bortolussi (1988) reported significant differences in Bedford ratings across four flight segments. However, the workload during hover was rated less than that during hover with a simultaneous communication task. Further, the scale was not sensitive to differences in neither control configurations nor combat countermeasure conditions. Lidderdale (1987) reported that post-flight ratings were very difficult for aircrews to make. Lastly, Vidulich (1991) questions whether the scale measures space capacity.

Data requirements - Roscoe (1984) suggested the use of short, well-defined flight tasks to enhance the reliability of subjective workload ratings. Harris, Hill, Lysaght, and Christ (1992) state that "some practice" is necessary to become familiar with the scale. They also suggest the use of nonparametric analysis technique since the Bedford is not an interval scale.

Thresholds - Minimum value is 1, maximum is 10.

Sources -

Corwin, W.H., Sandry-Garza, D.L., Biferno, M.H., Boucek, G.P., Logan, A.L., Jonsson, J.E., and Metalis, S.A. Assessment of crew workload measurement methods, techniques and procedures. Volume I - Process, methods, and results (WRDC-TR-89-7006). Wright-Patterson Air Force Base, OH; 1989.

Harris, R.M., Hill, S.G., Lysaght, R.J., and Christ, R.E. Handbook for operating the OWL & NEST technology (ARI Research Note 92-49). Alexandria, VA: United States Army Research Institute for the Behavioral and Social Sciences; 1992.

Lidderdale, I.G. Measurement of aircrew workload during low-level flight, practical assessment of pilot workload (AGARD-AG-282). Proceedings of NATO Advisory Group for Aerospace Research and Development (AGARD). Neuilly-sur-Seine, France: AGARD; 1987.

Roscoe, A.H. Assessing pilot workload in flight. Flight test techniques. Proceedings of NATO Advisory Group for Aerospace Research and Development (AGARD) (AGARD-CP-373). Neuilly-sur-Seine, France: AGARD; 1984.

Roscoe, A.H. In-flight assessment of workload using pilot ratings and heart rate. In A.H. Roscoe (Ed.) The practical assessment of pilot workload. AGARDograph No. 282 (pp. 78-82). Neuilly-sur-Seine, France: AGARD; 1987.

Tsang, P.S. and Johnson, W. Automation: Changes in cognitive demands and mental workload. Proceedings of the Fourth Symposium on Aviation Psychology. Columbus, OH: Ohio State University; 1987.

Vidulich, M.A. The Bedford Scale: Does it measure spare capacity? Proceedings of the 6th International Symposium on Aviation Psychology. 2, 1136-1141; 1991.

Vidulich, M.A. and Bortolussi, M.R. Control configuration study. Proceedings of the American Helicopter Society National Specialist's Meeting: Automation Applications for Rotorcraft; 1988.

Wainwright, W. Flight test evaluation of crew workload. In A.H. Roscoe (Ed.) The practical assessment of pilot workload. AGARDograph No. 282 (pp. 60-68). Neuilly-sur-Seine, France: AGARD; 1987.

3.2.4 Computerized Rapid Analysis of Workload

General description - The Computerized Rapid Analysis of Workload (CRAWL) is a computer program that helps designers predict workload in systems being designed. CRAWL inputs are mission timelines and task descriptions. Tasks are described in terms of cognitive, psychomotor, auditory, and visual demands.

Strengths and limitations - Bateman and Thompson (1986) reported increases in CRAWL ratings as task difficulty increased. Vickroy (1988) reported similar results as air turbulence increased.

Data requirements - The mission timeline must provide detailed second-by-second descriptions of the aircraft status.

Thresholds - Not stated.

Sources -

Bateman, R.P. and Thompson, M.W. Correlation of predicted workload with actual workload measured using the Subjective Workload Assessment Technique. Proceedings of the SAE AeroTech Conference; 1986.

Vickroy, S.C. Workload prediction validation study: The verification of CRAWL predictions; 1988.

3.2.5 Continuous Subjective Assessment of Workload

General description - The Continuous Subjective Assessment of Workload (C-SAW) requires subjects to provide ratings of 1 to 10 (corresponding to the Bedford Scale descriptors) while viewing a videotape of their flight immediately after landing. Computer prompts ratings at rates up to three seconds. A bar-chart or graph against the timeline is the output.

Strengths and limitations - Jensen (1995) stated that subjects could reliably provide ratings every three seconds. He reported C-SAW was sensitive to differences between a head-up display and a head-down display. C-SAW has high face validity but has not been formally validated.

Thresholds - The minimum is zero.

Source -

Jensen, S.E. Developing a flight workload profile using Continuous Subjective Assessment of Workload (C-SAW). Proceedings of the 21st Conference of the European Association for Aviation Psychology. Chapter 46, 1995.

3.2.6 Cooper-Harper Rating Scale

General description - The Cooper-Harper Rating Scale is a decision tree that uses adequacy for the task, aircraft characteristics, and demands on the pilot to rate handling qualities of an aircraft (see Figure 6).

Strengths and limitations - The Cooper-Harper Rating Scale is the current standard for evaluating aircraft handling qualities. It reflects differences in both performance and workload and is behaviorally anchored. It requires minimum training and a briefing guide has been developed (see Cooper and Harper, 1969, pp. 34-39). Cooper-Harper ratings have been sensitive to variations in controls, displays, and aircraft stability (Crabtree, 1975; Krebs and Wingert, 1976; Labacqz and Aiken, 1975; Schultz, Newell, and Whitbeck, 1970; Wierwille and Connor, 1983). Harper and Cooper (1984) describe a series of evaluations of the rating scale.

Connor and Wierwille (1983) reported significant increases in Cooper-Harper ratings as the levels of wind gust increased and/or as the aircraft pitch stability decreased. Ntuen, Park, Strickland, and Watson (1996) reported increases in Cooper-Harper ratings as instability in a compensatory tracking task increased. The highest ratings were for acceleration control; the lowest for position control; rate control was in the middle.

Data requirements - The scale provides ordinal data that must be analyzed accordingly. The Cooper-Harper scale should be used for workload assessment only if handling difficulty is the major determinant of workload. The task must be fully defined for a common reference.

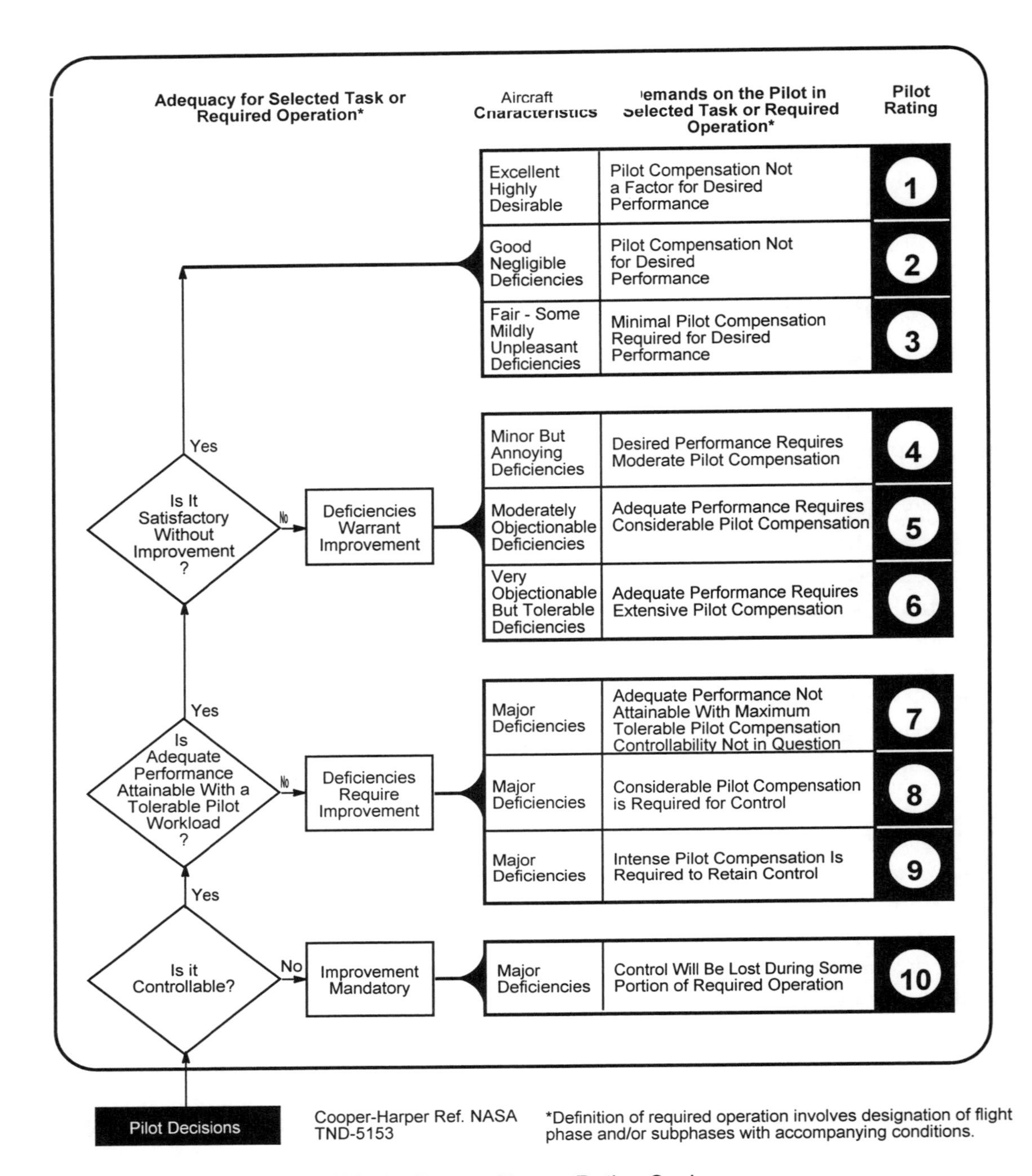

FIG. 6. Cooper-Harper Rating Scale

Thresholds - Ratings vary from 1 (excellent, highly desirable) to 10 (major deficiencies). Noninteger ratings are not allowed.

Sources -

Connor, S.A. and Wierwille, W.W. Comparative evaluation of twenty pilot workload assessment measures using a psychomotor task in a moving base aircraft simulator (Report 166457). Moffett Field, CA: NASA Ames Research Center; January 1983.

Cooper, G.E. and Harper, R.P. The use of pilot rating in the evaluation of aircraft handling qualities (AGARD Report 567). London: Technical Editing and Reproduction Ltd.; April 1969.

Crabtree, M.S. Human factors evaluation of several control system configurations, including workload sharing with force wheel steering during approach and flare (AFFDL-TR-75-43). Wright-Patterson Air Force Base, OH: Flight Dynamics Laboratory; April 1975.

Harper, R.P. and Cooper, G.E. Handling qualities and pilot evaluation. AIAA, AHS, ASEE, Aircraft Design Systems and Operations meeting, AIAA Paper 84-2442; 1984.

Krebs, M.J. and Wingert, J.W. Use of the oculometer in pilot workload measurement (NASA CR-144951). Washington, DC: National Aeronautics and Space Administration; February 1976.

Lebacqz, J.V. and Aiken, E.W. A flight investigation of control, display, and guidance requirements for decelerating descending VTOL instrument transitions using the X-22A variable stability aircraft (AK-5336-F-1). Buffalo, NY: Calspan Corporation; September 1975.

Ntuen, C.A., Park, E., Strickland, D. and Watson, A.R. A frizzy model for workload assessment in complex task situations. IEEE 0-8186-7493, August 1996, pp. 101-107.

Schultz, W.C., Newell, F.D., and Whitbeck, R.F. A study of relationships between aircraft system performance and pilot ratings. Proceedings of the Sixth Annual NASA University Conference on Manual Control, Wright-Patterson Air Force Base, OH. 339-340; April 1970.

Wierwille, W.W. and Connor, S.A. Evaluation of 20 workload measures using a psychomotor task in a moving-base aircraft simulator. Human Factors. 25(1), 1-16, 1983.

3.2.7 Crew Status Survey

General description - The original Crew Status Survey was developed by Pearson and Byars (1956) and contained 20 statements describing fatigue states. The staff of the Air Force School of Aerospace Medicine Crew Performance Branch, principally Storm and Parke, updated the original survey. They selected the statements anchoring the points on the fatigue scale of the survey through iterative presentations of drafts of the survey to aircrew members. The structure of the fatigue scale was somewhat cumbersome, since the dimensions of workload, temporal demand, system demand, system management, danger, and acceptability were combined on one scale. However, the fatigue scale was simple enough to be well received by operational crews. The fatigue scale of the survey was shortened to seven statements and subsequently tested for sensitivity to fatigue as well as for test/retest reliability (Miller and Narvaez, 1986). Finally, a seven-point workload scale was added. The current Crew Status Survey (see Figure 7) provides measures of self-reported fatigue and workload as well as space for general comments. Ames and George (1993) are modifying the workload scale to enhance reliability. Their scale descriptors are:

(1) Nothing to do; No system demands.
(2) Light activity; Minimum demands.
(3) Moderate activity; Easily managed; Considerable spare time.
(4) Busy; challenging but manageable; Adequate time available.
(5) Very busy; Demanding to manage; Barely enough time.
(6) Extremely busy; Very difficult; Non-essential tasks postponed.
(7) Overloaded; System unmanageable; Important tasks undone; unsafe. (p. 4).

NAME	DATE AND TIME

SUBJECT FATIGUE

(Circle the number of the statement which describes how you feel RIGHT NOW.)

1	Fully Alert, Wide Awake; Extremely Peppy
2	Very Lively; Responsive, But Not at Peak
3	Okay; Somewhat Fresh
4	A Little Tired; Less Than Fresh
5	Moderately Tired; Let Down
6	Extremely Tired; Very Difficult to Concentrate
7	Completely Exhausted; Unable to Function Effectively; Ready to Drop

COMMENTS

WORKLOAD ESTIMATE

(Circle the number of the statement which describes the MAXIMUM workload you experienced during the past work period. Put an X over the number of the statement which best describes the AVERAGE workload you experienced during the past work period.)

1	Nothing to do; No System Demands
2	Little to do; Minimum System Demands
3	Active Involvement Required, But Easy to Keep Up
4	Challenging, But Manageable
5	Extremely Busy; Barely Able to Keep Up
6	Too Much to do; Overloaded; Postponing Some Tasks
7	Unmanageable; Potentially Dangerous; Unacceptable

COMMENTS

SAM FORM **202** **CREW STATUS SURVEY**
APR 81

PREVIOUS EDITION WILL BE USED

FIG. 7. Crew Status Survey

Strengths and limitations - These scales have been found to be sensitive to changes in task demand and fatigue but are independent of each other (Courtright, Frankenfeld, and Rokicki, 1986). Storm and Parke (1987) used the Crew Status Survey to assess the effects of temazepam on FB-111A crewmembers. The effect of the drug was not significant. The effect of performing the mission was, however. Specifically, the fatigue ratings were higher at the end than at the beginning of a mission. Gawron, et al. (1988) analyzed Crew Status Survey ratings made at four times during each flight. They found a significant segment effect on fatigue and workload. Fatigue ratings increased over the course of the flight (preflight = 1.14, predrop = 1.47, postdrop = 1.43, and postflight = 1.56). Workload ratings were highest around a simulated air drop (preflight = 1.05, predrop = 2.86, postdrop = 2.52, and postflight = 1.11).

George, Nordeen, and Thurmond (1991) collected workload ratings from Combat Talon II aircrew members during arctic deployment. None of the median ratings were greater than four. However, level 5 ratings occurred for navigators during airdrops and self-contained approach run-ins. These authors also used the Crew Status Survey workload scale during terrain-following training flights on Combat Talon II. Pilots and copilots gave a median rating of 7. The ratings were used to identify major crewstation deficiencies.

However, George and Hollis (1991) reported confusion between adjacent categories at the high workload end of the Crew Status Survey. They also found adequate ordinal properties for the scale but very large variance in most order-of-merit tables.

Data requirements - Although the Crew Status Survey is printed on card stock, subjects find it difficult to fill in the rating scale during high workload periods. Further, sorting (for example, by the time completed) the completed card-stock ratings after the flight is also difficult and not error free. A larger character-size version of the survey has been included on flight cards at the Air Force Flight Test Center. Verbal ratings prompted by the experimenter work well if: (1) subjects can quickly scan a card-stock copy of the rating scale to verify the meaning of a rating and (2) subjects are not performing a conflicting verbal task. Each scale can be used independently.

Thresholds - 1 to 7 for subjective fatigue; 1 to 7 for workload (see Figure 7).

Sources -

Ames, L.L. and George, E.J. Revision and verification of a seven-point workload estimate scale (AFFTC-TIM-93-01). Edwards Air Force Base, CA: Air Force Flight Test Center, 1993.

Courtright, J.F., Frankenfeld, C.A., and Rokicki, S.M. The independence of ratings of workload and fatigue. Paper presented at the Human Factors Society 30th Annual Meeting, Dayton, Ohio; 1986.

Gawron, V.J., Schiflett, S.G., Miller, J., Ball, J., Slater, T., Parker, F., Lloyd, M., Travale, D., and Spicuzza, R.J. The effect of pyridostigmine bromide on inflight aircrew performance (USAFSAM-TR-87-24). Brooks Air Force Base, TX: School of Aerospace Medicine; January 1988.

George, E. and Hollis, S. Scale validation in flight test. Edwards Air Force Base, CA: Flight Test Center; December 1991.

George, E.J., Nordeen, M., and Thurmond, D. Combat Talon II human factors assessment (AFFTC TR 90-36). Edwards Air Force Base, CA: Flight Test Center; November 1991.

Miller, J.C. and Narvaez, A. A comparison of two subjective fatigue checklists. Proceedings of the 10th Psychology in the DoD Symposium. Colorado Springs, CO: United States Air Force Academy, 514-518; 1986.

Pearson, R.G. and Byars, G.E. The development and validation of a checklist for measuring subjective fatigue (TR-56-115). Brooks Air Force Base, TX: School of Aerospace Medicine; 1956.

Storm, W.F. and Parke, R.C. FB-111A aircrew use of temazepam during surge operations. Proceedings of the NATO Advisory Group for Aerospace Research and Development (AGARD) Biochemical Enhancement of Performance Conference (Paper number 415, p. 12-1 to 12-12). Neuilly-sur-Seine, France: AGARD; 1987.

3.2.8 Dynamic Workload Scale

General description - The Dynamic Workload Scale is a seven-point workload scale (see Figure 8) developed as a tool for aircraft certification. It has been used extensively by Airbus Industries.

Strengths and limitations - Speyer, Fort, Fouillot, and Bloomberg (1987) reported high concordance between pilot and observer ratings as well as sensitivity to workload increases.

Data requirements - Dynamic Workload Scale ratings must be given by both a pilot and an observer-pilot. The pilot is cued to make a rating; the observer gives a rating whenever workload changes or five minutes have passed.

Thresholds - Two is minimum workload; eight, maximum workload.

Source -

Speyer, J., Fort, A., Fouillot, J., and Bloomberg, R. Assessing pilot workload for minimum crew certification. In A.H. Roscoe (Ed.) The practical assessment of pilot workload. AGARDograph Number 282 (pp. 90-115). Neuilly-sur-Seine, France: AGARD; 1987.

3.2.9 Equal-Appearing Intervals

General description - Subjects rate the workload in one of several categories using the assumption that each category is equidistant from adjacent categories.

Strengths and limitations - Hicks and Wierwille (1979) reported sensitivity to task difficulty in a driving simulator. Masline (1986) reported comparable results with the magnitude estimates and Subjective Workload Assessment Technique (SWAT) ratings but greater ease of administration. Masline, however, warned of rater bias.

Workload Assessment		**CRITERIA**			**Appreciation**
		Reserve Capacity	**Interruptions**	**Effort or Stress**	
Light	2	Ample			Very Acceptable
Moderate	3	Adequate	Some		Well Acceptable
Fair	4	Sufficient	Recurring	Not Undue	Acceptable
High	5	Reduced	Repetitive	Marked	High but Acceptable
Heavy	6	Little	Frequent	Significant	Just Acceptable
Extreme	7	None	Continuous	Acute	Not Acceptable Continuously
Supreme	8	Impairment	Impairment	Impairment	Not Acceptable Instantaneously

from Lysaght, et al. (1989) p. 108

FIG. 8. Dynamic Workload Scale

Data requirements - Equal intervals must be clearly defined.
Thresholds - Not stated.
Sources -

Hicks, T.G. and Wierwille, W.W. Comparison of five mental workload assessment procedures in a moving-base driving simulator. Human Factors. 21, 129-143; 1979.
Masline, P.J. A comparison of the sensitivity of interval scale psychometric techniques in the assessment of subjective mental workload. Unpublished masters thesis, University of Dayton, Dayton, OH; 1986.

3.2.10 Finegold Workload Rating Scale

General description - The Finegold Workload Rating Scale has five subscales (see Figure 9). It was developed to evaluate workload at each crewstation aboard the AC-130H Gunship.

Strengths and limitations - Finegold, Lawless, Simons, Dunleavy, and Johnson (1986) reported lower ratings associated with cruise than with engagement or threat segments. Analysis of the subscales indicated that time stress was rated differently at each crewstation. Lozano (1989) replicated the Finegold, et al. (1986) test using the AC-1304 Gunship, again, ratings on subscales varied by crew position. George (1994) is replicating both studies using the current version of the AC-130U Gunship.

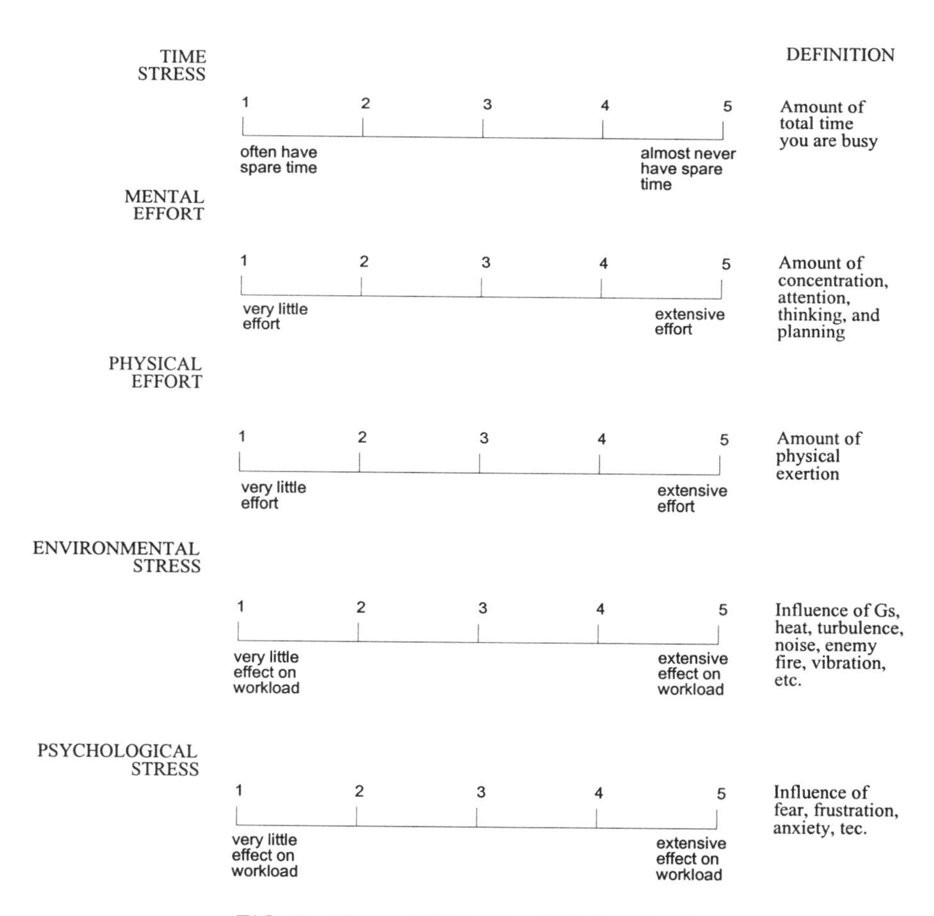

FIG. 9. Finegold Workload Rating Scale

Data requirements - Average individual subscales as well as the complete workload Rating Scale scores.

Thresholds - 1 for low workload and 5 for high workload.

Sources -

Finegold, L.S., Lawless, M.T., Simons, J.L., Dunleavy, A.O., and Johnson, J. Estimating crew performance in advanced systems. Volume II: Application to future gunships Appendix B: Results of data analysis for AC-13OH and hypothetical AC-130H (RP). Edwards Air Force Base, CA: Air Force Flight Test Center; October 1986.

George, E.J. AC-130U gunship workload evaluation (C4654-3-501). Edwards AFB, CA: Air Force Flight Test Center; April 1994.

Lozano, M.L. Human engineering test report for the AC-130U gunship (NA-88-1805). Los Angeles, CA: Rockwell International; January 1989.

3.2.11 Flight Workload Questionnaire

General description - The Flight Workload Questionnaire is a four-item, behaviorally anchored rating scale. The items and the end points of the rating scales are: workload category (low to very high), fraction of time busy (seldom have much to do to fully occupied at all times), how hard had to think (minimal thinking to a great deal of thinking), and how felt (relaxing to very stressful).

Strengths and limitations - The questionnaire is sensitive to differences in experience and ability. For example, Stein (1984) found significant differences in the flight workload ratings between experienced and novice pilots. Specifically, experienced pilots rated their workload during an air transport flight lower than novice pilots did. However, Stein also found great redundancy in the value of the ratings given for the four questionnaire items. This suggests that the questionnaire may evoke a response bias. The questionnaire provides a measure of overall workload but cannot differentiate between flight segments and/or events.

Data requirements - Not stated.

Thresholds - Not stated.

Source -

Stein, E.S. The measurement of pilot performance: A master-journeyman approach (DOT/FAA/CT-83/15). Atlantic City, NJ: Federal Aviation Administration Technical Center; May 1984.

3.2.12 Hart and Bortolussi Rating Scale

General description - Hart and Bortolussi (1984) used a single rating scale to estimate workload. The scale units were 1 to 100 with 1 being low workload and 100 being high workload.

Strengths and limitations - The workload ratings significantly varied across flight segments with takeoff and landing having higher workload than climb or cruise. The workload ratings were significantly correlated to ratings of stress (+0.75) and effort (+0.68). These results were based on data from 12 instrument-rated pilots reviewing a list of 163 events.

Moray, Dessouky, Kijowski, and Adapathya (1991) used the same rating scale but numbered the scale from 1 to 10 rather than from 1 to 100. This measure was significantly related to time pressure but not to knowledge or their interaction.

Data requirements - The subjects need only the end points of the scale.
Thresholds - 1 = low workload, 100 = high workload.
Sources -

Hart, S.A. and Bortolussi, M.R. Pilot errors as a source of workload. Human Factors. 25(5), 545-556; 1984.

Moray, N., Dessouky, M.I., Kijowski, B.A., and Adapathya, R.S. Strategic behavior, workload, and performance in task scheduling. Human Factors. 33(6), 607-629; 1991.

3.2.13 Hart and Hauser Rating Scale

General description - Hart and Hauser (1987) used a six-item rating scale (see Figure 10) to measure workload during a nine-hour flight. The items and their scales were: stress (completely relaxed to extremely tense), mental/sensory effort (very low to very high), fatigue (wide awake to worn out), time pressure (none to very rushed), overall workload (very low to very high), and performance (completely unsatisfactory to completely satisfactory). Subjects were instructed to mark the scale position that represented their experience.

Strengths and limitations - The scale was developed for use in flight. In the initial study, Hart and Hauser (1987) asked subjects to complete the questionnaire at the end of each of seven flight segments. They reported significant segment effects in the seven-hour flight. Specifically, stress, mental/sensory effort, and time pressure were lowest during a data-recording segment. There was a sharp increase in rated fatigue after the start of the data-recording segment. Overall, the aircraft commander rated workload as higher than by the copilot did. Finally, performance received the same ratings throughout the flight.

Data requirements - The scale is simple to use but requires a stiff writing surface and minimal turbulence.

Thresholds - Not stated.

Source -

Hart, S.G. and Hauser, J.R. Inflight application of three pilot workload measurement techniques. Aviation, Space, and Environmental Medicine. 58, 402-410; 1987.

Stress

Completely Relaxed ________________________ Extremely Tense

Mental/Sensory Effort

Very Low ________________________ Very High

Fatigue

Wide Awake ________________________ Worn Out

Time Pressure

None ________________________ Very Rushed

Overall Workload

Very Low ________________________ Very High

Performance

Completely Unsatisfactory ________________________ Completely Satisfactory

FIG. 10. Hart and Hauser Rating Scale

3.2.14 Honeywell Cooper-Harper Rating Scale

General description - This rating scale (see Figure 11) uses a decision-tree structure for assessing overall task workload.

Strengths and limitations - The Honeywell Cooper-Harper Rating Scale was developed by Wolf (1978) to assess overall task workload. North, Stackhouse, and Graffunder (1979) used the scale to assess workload associated with various Vertical Take-Off and Landing (VTOL) aircraft displays. For the small subset of conditions analyzed, the scale ratings correlated well with performance.

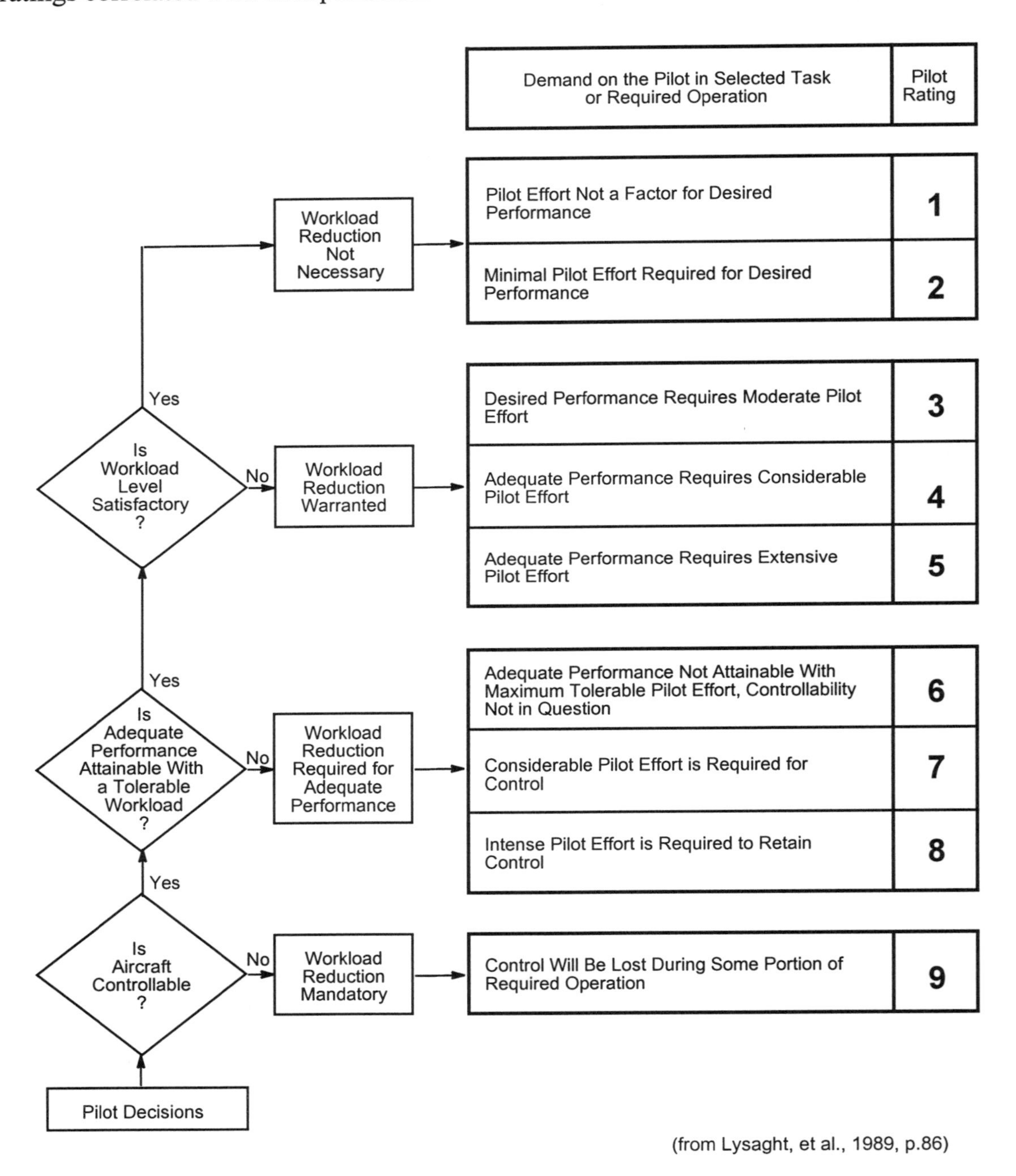

FIG. 11. Honeywell Cooper-Harper Rating Scale

Data requirements - Subjects must answer three questions related to task performance. The ratings are ordinal and must be treated as such in subsequent analyses.

Thresholds - Minimum is 1, maximum is 9.

Sources -

Lysaght, R.J., Hill, S.G., Dick, A.O., Plamondon, B.D., Linton, P.M., Wierwille, W.W., Zaklad, A.L., Bittner, A.C., and Wherry, R.J. Operator workload: comprehensive review and evaluation of operator workload methodologies (Technical Report 851). Alexandria, VA: Army Research Institute for the Behavioral and Social Sciences; June 1989.

North, R.A., Stackhouse, S.P., and Graffunder, K. Performance, physiological and oculometer evaluations of VTOL landing displays (NASA Contractor Report 3171). Hampton, VA: NASA Langley Research Center; 1979.

Wolf, J.D. Crew workload assessment: Development of a measure of operator workload (AFFDL-TR-78-165). Wright-Patterson AFB, OH: Air Force Flight Dynamics Laboratory; 1978.

3.2.15 Magnitude Estimation

General description - Subjects are required to estimate workload numerically in relation to a standard.

Strengths and limitations - Borg (1978) successfully used this method for evaluating workload. Helm and Heimstra (1981) reported a high correlation between workload estimates and task difficulty. Masline (1986) reported sensitivity comparable to estimates from the equal-appearing intervals method and SWAT. Gopher and Braune (1984), however, found a low correlation between workload estimates and reaction-time performance. In contrast, Kramer, Sirevaag, and Braune (1987) reported good correspondence to performance in a fixed-based flight simulator. Hart and Staveland (1988) suggest that the presence of a standard enhances interrater reliability. O'Donnell and Eggemeier (1986), however, warned that subjects may be unable to retain an accurate memory of the standard over the course of an experiment.

Data requirements - A standard must be well defined.

Thresholds - Not stated.

Sources -

Borg, C.G. Subjective aspects of physical and mental load. Ergonomics. 21, 215-220; 1978.

Gopher, D. and Braune, R. On the psychophysics of workload: Why bother with subjective measures? Human Factors. 26, 519-532; 1984.

Hart, S.G. and Staveland, L.E. Development of NASA-TLX (Task Load Index): Results of empirical and theoretical research. In P.A. Hancock and N. Meshkati (Eds) Human Mental Workload. Amsterdam: Elsevier; 1988.

Helm, W. and Heimstra, N.W. The relative efficiency of psychometric measures of task difficulty and task performance in predictive task performance (Report No. HFL-81-5). Vermillion, S.D.: University of South Dakota, Psychology Department, Human Factors Laboratory; 1981.

Kramer, A.F., Sirevaag, E.J., and Braune, R. A psychophysical assessment of operator workload during simulated flight missions. Human Factors. 29, 145-160; 1987.

Masline, P.J. A comparison of the sensitivity of interval scale psychometric techniques in the assessment of subjective workload. Unpublished masters thesis. Dayton, OH: University of Dayton; 1986.

O'Donnell, R.D. and Eggemeier, F.T. Workload assessment methodology. In K.R. Boff, L. Kaufman, and J. Thomas (Eds.) Handbook of perception and human performance. Vol. 2, Cognitive processes and performance. New York: Wiley; 1986.

3.2.16 McCracken-Aldrich Technique

General description - The McCracken-Aldrich Technique was developed to identify workload associated with flight control, flight support, and mission-related activities (McCracken and Aldrich, 1984).

Strength and limitations - The technique may require months of preparation to use. It has been useful in assessing workload in early system design stages.

Data requirements - A mission must be decomposed into segments, functions, and performance elements (e.g., tasks). Subjective Matter Experts rate workload (from 1 to 7) for each performance element. A FORTRAN programmer is required to generate the resulting scenario timeline.

Thresholds - Not Stated.

Source -

McCracken, J.H. and Aldrich, T.B. Analysis of selected LHX mission functions: implications for operator workload and system automation goals (TNA ASI 479-24-84). Fort Rucker, AL: Anacapa Sciences; 1984.

3.2.17 McDonnell Rating Scale

General description - The McDonnell rating scale (see Figure 12) is a ten-point scale requiring a pilot to rate workload based on the attentional demands of a task.

Strengths and limitations - Van de Graaff (1987) reported significant differences in workload among various flight approach segments and crew conditions. Intersubject variability among McDonnell ratings was less than that among SWAT ratings.

Data requirements - Not stated.

Thresholds - Not stated.

Sources -

McDonnell, J.D. Pilot rating techniques for the estimation and evaluation of handling qualities (AFFDL-TR-68-76). Wright-Patterson Air Force Base, TX: Air Force Flight Dynamics Laboratory; 1968.

van de Graaff, R.C. An in-flight investigation of workload assessment techniques for civil aircraft operations, (NLR-TR-87119U). Amsterdam, the Netherlands: National Aerospace Laboratory; 1987.

3.2.18 Mission Operability Assessment Technique

General description - The Mission Operability Assessment Technique includes two four-point ordinal rating scales, one for pilot workload, the other for technical effectiveness (see Table 21). Subjects rate both pilot workload and technical effectiveness for each subsystem identified in the task analysis of the aircraft.

<table>
<tr><td rowspan="9">Controllable
Capable of being controlled or managed in context of mission, with available pilot attention</td><td rowspan="6">Acceptable
May have deficiencies which warrant improvement, but adequate for mission. Pilot compensation, if required to achieve acceptable performance, is feasible.</td><td rowspan="3">Satisfactory
Meets all requirements and expectations; good enough without improvement. Clearly adequate for mission.</td><td>Excellent, Highly desirable</td><td>A1</td></tr>
<tr><td>Good, pleasant, well behaved</td><td>A2</td></tr>
<tr><td>Fair. Some mildly unpleasant characteristics. Good enough for mission without improvement.</td><td>A3</td></tr>
<tr><td rowspan="3">Unsatisfactory
Reluctantly acceptable. Deficiencies which warrant improvement. Performance adequate for mission with feasible pilot compensation.</td><td>Some minor but annoying deficiencies. Improvement is requested. Effect on performance is easily compensated for by pilot.</td><td>A4</td></tr>
<tr><td>Moderately objectionable deficiencies. Improvement is needed. Reasonable performance requires considerable pilot compensation.</td><td>A5</td></tr>
<tr><td>Very objectionable deficiencies. Major improvements are needed. Requires best available pilot compensation to achieve acceptable performance.</td><td>A6</td></tr>
<tr><td rowspan="3">Unacceptable
Deficiencies which require mandatory improvement. Inadequate performance for mission, even with maximum feasible pilot compensation.</td><td rowspan="3"></td><td>Major deficiencies which require mandatory improvement for acceptance. Controllable. Performance inadequate for mission, or pilot compensation required for minimum acceptable performance in mission is too high.</td><td>U7</td></tr>
<tr><td>Controllable with difficulty. Requires substantial pilot skill and attention to retain control and continue mission.</td><td>U8</td></tr>
<tr><td>Marginally controllable in mission. Requires maximum available pilot skill and attention to retain control</td><td>U9</td></tr>
<tr><td colspan="3">Uncontrollable
Control will be lost during some portion of the mission.</td><td>Uncontrollable in Mission</td><td>U10</td></tr>
</table>

from McDonnell, 1968, p. 7

FIG. 12. McDonnell Rating Scale

TABLE 21.
Mission Operability Assessment Technique Pilot Workload and Subsystem Technical Effectiveness Rating Scales

Pilot Workload

1. The pilot workload (PW)/compensation (C)/interference (I) required to perform the designated task is *extreme*. This is a *poor* rating on the PW/C/I dimension.
2. The pilot workload/compensation/interference required to perform the designated task is *high*. This is a *fair* rating on the PW/C/I dimension.
3. The pilot workload/compensation/interference required to perform the designated task is *moderate*. This is a *good* rating on the PW/C/I dimension.
4. The pilot workload/compensation/interference required to perform the designated task is *low*. This is an *excellent* rating on the PW/C/I dimension.

Subsystem Technical Effectiveness

1. The technical effectiveness of the required subsystem is *inadequate* for performing the designated task. Considerable redesign is necessary to attain task requirements. This is a *poor* rating on the subsystem technical effectiveness scale.
2. The technical effectiveness of the required subsystem is *adequate* for performing the designated task. Some redesign is necessary to attain task requirements. This is a *fair* rating on the subsystem technical effectiveness scale.
3. The technical effectiveness of the required subsystem *enhances individual task performance*. No redesign is necessary to attain task requirements. This is a *good* rating on the subsystem technical effectiveness scale.
4. The technical effectiveness of the required subsystem *allows for the integration of multiple tasks*. No redesign is necessary to attain task requirements. This is an *excellent* rating on the subsystem effectiveness scale. (O'Donnell and Eggemeier, 1986, p. 42-16)

Strengths and limitations - Interrater reliabilities are high for most but not all tasks (Donnell, 1979; Donnell, Adelman, and Patterson, 1981; Donnell and O'Connor, 1978).

Data requirements - Conjoint measurement techniques are applied to the individual pilot workload and subsystem technical effectiveness ratings to develop an overall interval scale of systems capability.

Thresholds - Not stated.

Sources -

Donnell, M.L. An application of decision-analytic techniques to the test and evaluation of a major air system Phase III (TR-PR-79-6-91). McLean, VA: Decisions and Designs; May 1979.

Donnell, M.L., Adelman, L, and Patterson, J.F. A systems operability measurement algorithm (SOMA): application, validation, and extensions (TR-81-11-156). McLean, VA: Decisions and Designs; April 1981.

Donnell, M.L. and O'Connor, M.F. The application of decision analytic techniques to the test and evaluation phase of the acquisition of a major air system Phase II (TR-78-3-25). McLean, VA: Decisions and Designs; April 1978.

O'Donnell, R.D. and Eggemeier, F.T. Workload assessment methodology. In K.R. Boff, L. Kaufman, and J.P. Thomas (Eds.) Handbook of perception and human performance. New York: John Wiley, 1986.

3.2.19 Modified Cooper-Harper Rating Scale

General description - Wierwille and Casali (1983) noted that the Cooper-Harper scale represented a combined handling-qualities/workload rating scale. They found that it was sensitive to psychomotor demands on an operator, especially for aircraft handling qualities. They wanted to develop an equally useful scale for the estimation of workload associated with cognitive functions, such as "perception, monitoring, evaluation, communications, and problem solving." The Cooper-Harper scale terminology was not suited to this purpose. A modified Cooper-Harper rating scale (see Figure 13) was developed to "increase the range of applicability to situations commonly found in modern systems." Modifications included: (1) changing the rating-scale end points to very easy and impossible, (2) asking the pilot to rate mental workload level rather than controllability, and (3) emphasizing difficulty rather than deficiencies. In addition, Wierwille and Casali (1983) defined mental effort as "minimal" in rating 1, while mental effort is not defined as minimal until rating 3 in the original Cooper-Harper scale. Further, adequate performance begins at rating 3 in the modified Cooper-Harper but at rating 5 in the original scale.

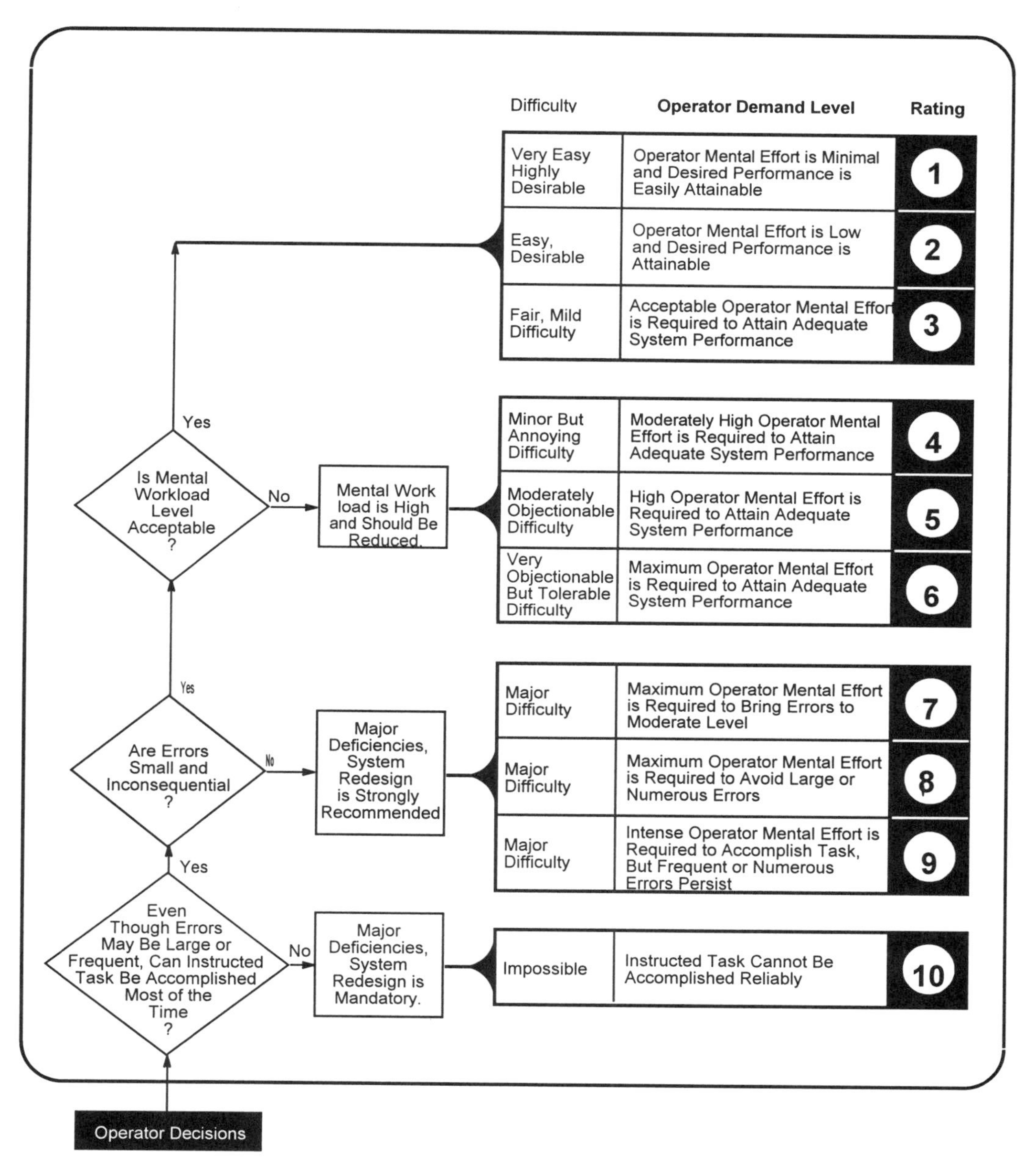

FIG. 13. Modified Cooper-Harper Rating Scale

Strengths and limitations - Investigations were conducted to assess the modified Cooper-Harper scale. They focused on perception (e.g., aircraft engine instruments out of limits during simulated flight), cognition (e.g., arithmetic problem solving during simulated flight), and communications (e.g., detection of, comprehension of, and response to own aircraft call sign during simulated flight).

The modified Cooper-Harper is sensitive to various types of workloads. For example, Casali and Wierwille (1983) reported that modified Cooper-Harper ratings increased as the communication load increased. Wierwille, Rahimi, and Casali (1985) reported significant increase in workload as navigation load increased. Casali and Wierwille (1984) reported significant increases in ratings as the number of danger conditions increased. Skipper, Rieger, and Wierwille (1986) reported significant increases in ratings in both high communication and high navigation loads. Wolf (1978) reported the highest workload ratings in the highest workload flight condition (i.e., high wind gust and poor handling qualities).

Bittner, Byers, Hill, Zaklad, and Christ (1989) reported reliable differences between mission segments in a mobile air defense system. Byers, Bittner, Hill, Zaklad, and Christ (1988) reported reliable differences between crew positions in a remotely piloted vehicle system. These results suggested that the modified Cooper-Harper scale is a valid, statistically reliable indicator of overall mental workload. However, it carries with it the underlying assumptions that high workload is the only determinant of the need for changing the control/display configuration. Wierwille, Casali, Connor, and Rahimi (1985) concluded that the modified Cooper-Harper Rating Scale provided consistent and sensitive ratings of workload across a range of tasks. Wierwille, Skipper, and Rieger (1985) reported the best consistency and sensitivity with the modified Cooper-Harper from five alternatives tests. Warr, Colle, and Reid (1986) reported that the modified Cooper-Harper Ratings were as sensitive to task difficulty as SWAT ratings. Kilmer, Knapp, Burdsal, Borresen, Bateman, and Malzahn (1988), however, reported that the modified Cooper-Harper rating scale was less sensitive than SWAT ratings to changes in tracking task difficulties. Hill, Iavecchia, Byers, Bittner, Zaklad, and Christ (1992) reported that the modified Cooper-Harper scale was not as sensitive or as operator accepted as the NASA TLX or the overall workload scale.

Papa and Stoliker (1988) tailored the modified Cooper-Harper rating scale to evaluate the Low Altitude Navigation and Targeting Infrared System for Night (LANTIRN) on an F-16 aircraft.

Data requirements - Wierwille and Casali (1983) recommend the use of the modified Cooper-Harper in experiments where overall mental workload is to be assessed. They emphasize the importance of proper instructions to the subjects. Since the scale was designed for use in experimental situations, it may not be appropriate to situations requiring an absolute diagnosis of a subsystem. Harris, Hill, Lysaght, and Christ (1992) recommend the use of non-parametric analysis techniques since the modified Cooper-Harper rating scale is not an interval scale.

Thresholds - Not stated.

Sources -

Bittner, A.C., Byers, J.C., Hill, S.G., Zaklad, A.L., and Christ, R.E. Generic workload ratings of a mobile air defense system (LOS-F-H). Proceedings of the 33rd Annual

Meeting of the Human Factors Society (pp. 1476-1480). Santa Monica, CA: Human Factors Society; 1989.

Byers, J.C., Bittner, A.C., Hill, S.G., Zaklad, A.L., and Christ, R.E. Workload assessment of a remotely piloted vehicle (RPV) system. Proceedings of the 32nd Annual Meeting of the Human Factors Society (pp. 1145-1149). Santa Monica, CA: Human Factors Society; 1988.

Casali, J.G. and Wierwille, W.W. A comparison of rating scale, secondary task, physiological, and primary-task workload estimation techniques in a simulated flight task emphasizing communications load. Human Factors. 25, 623-642; 1983.

Casali, J.G. and Wierwille, W.W. On the comparison of pilot perceptual workload: A comparison of assessment techniques addressing sensitivity and intrusion issues. Ergonomics. 27, 1033-1050; 1984.

Harris, R.M., Hill, S.G., Lysaght, R.J., and Christ, R.E. Handbook for operating the OWLKNEST technology (ARI Research Note 92-49). Alexandria, VA: United States Army Research institute for the Behavioral and Social Sciences; 1992.

Hill, S.G., Iavecchia, H.P., Byers, J.C., Bittner, A.C., Zaklad, A.L., and Christ, R.E. comparison of four subjective workload rating scales. Human Factors. 34, 429-439; 1992.

Kilmer, K.J., Knapp, R., Burdsal, C., Borresen, R., Bateman, R., and Malzahn, D. Techniques of subjective assessment: A comparison of the SWAT and modified Cooper-Harper scale. Proceedings of the Human Factors Society 32nd Annual Meeting. 155-159; 1988.

Papa, R.M. and Stoliker, J.R. Pilot workload assessment: a flight test approach. Washington, DC: American Institute of Aeronautics and Astronautics, 88-2105, 1988.

Skipper, J.H., Rieger, C.A., and Wierwille, W.W. Evaluation of decision-tree rating scales for mental workload estimation. Ergonomics. 29, 585-599; 1986.

Warr, D., Colle, H. and Reid, G. A comparative evaluation of two subjective workload measures: The subjective workload assessment technique and the modified Cooper-Harper scale. Paper presented at the Symposium on Psychology in Department of Defense. Colorado Springs, CO: US Air Force Academy; 1986.

Wierwille, W.W. and Casali, J.G. A validated rating scale for global mental workload measurement applications. Proceedings of the 27th Annual Meeting of the Human Factors Society. 129-133. Santa Monica, CA: Human Factors Society; 1983.

Wierwille, W.W., Casali, J.G., Connor, S.A., and Rahimi, M. Evaluation of the sensitivity and intrusion of mental workload estimation techniques. In W. Romer (Ed.) Advances in man-machine systems research. Volume 2 (pp. 51-127). Greenwich, CT: J.A.I. Press; 1985.

Wierwille, W.W., Rahimi, M., and Casali, J.G. Evaluation of 16 measures of mental workload using a simulated flight task emphasizing mediational activity. Human Factors. 27(5), 489-502; 1985.

Wierwille, W.W., Skipper, J. and Reiger, C. Decision tree rating scales for workload estimation: theme and variations (N85-11544), Blacksburg, VA: Vehicle Simulation Laboratory; 1985.

Wolf, J.D. Crew workload assessment: Development of a measure of operator workload (AFFDL-TR-78-165). Wright-Patterson AFB, OH: Air Force Flight Dynamics Laboratory; December 1978.

3.2.20 *Multi-Descriptor Scale*

General description - The Multi-descriptor (MD) scale is composed of six descriptors: 1) attentional demand, 2) error level, 3) difficulty, 4) task complexity, 5) mental workload, and 6) stress level. Each descriptor is rated after a task. The MD score is the average of the six descriptor ratings.

Strengths and limitations - Wierwille, Rahimi, and Casali (1985) reported the MD scores were not sensitive to variations in difficulty of mathematical calculations performed during a simulated flight task.

Data requirements - The six rating scales must be presented to the subject after a flight and the average of the resultant ratings calculated.

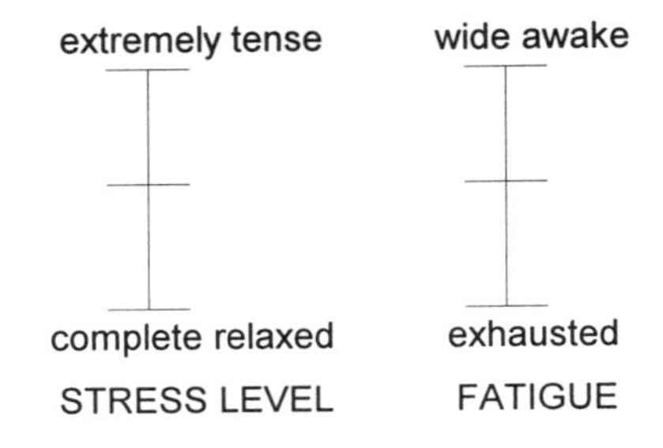

Source -

Wierwille, W.W., Rahimi, M., and Casali, J.G. Evaluation of 16 measures of mental workload using a simulated flight task emphasizing mediational activity. Human Factors. 27(5), 489-502; 1985.

3.2.21 *Multidimensional Rating Scale*

General description – The Multidimensional Rating Scale is composed of eight bipolar scales (see Table 22). Subjects are asked to draw a horizontal line on the scale to indicate their rating.

TABLE 22.
Multidimensional Rating Scale

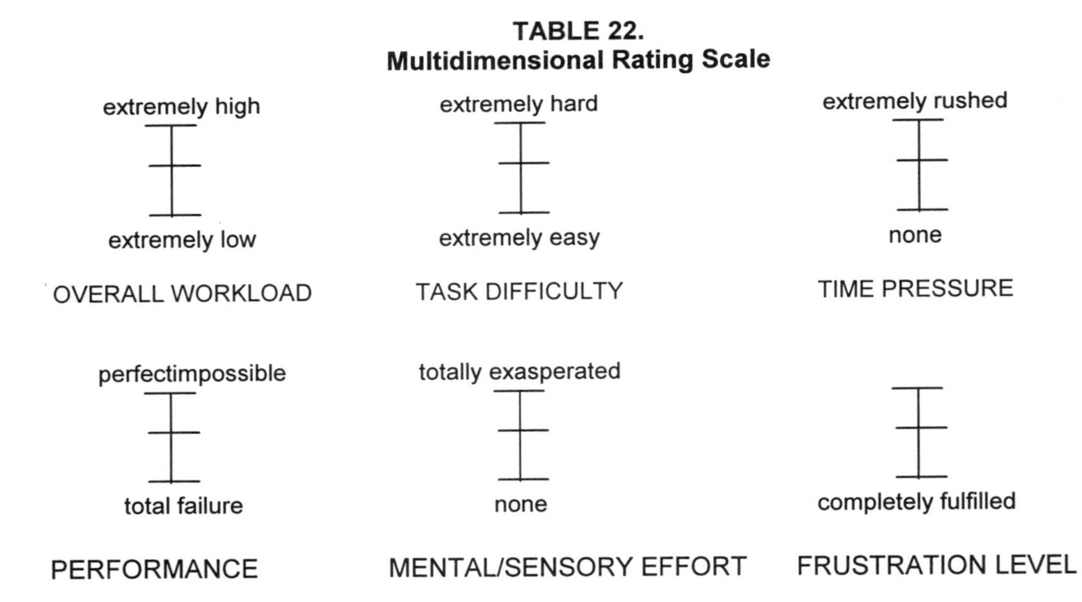

Strengths and limitations - Damos (1985) reported high correlations among several of the subscales (+0.82 between overall workload and task difficulty single-task condition, +0.73 task variations. The time pressure and overall workload scales were also significantly associated with task by pacing condition interactions. The mental/sensory effort scale was significantly associated with a task by behavior pattern interaction.

Data requirements – The vertical line in each scale must be 100 ml long. The rater must measure the distance from the bottom of the scale to the subject's horizontal line to determine the rating.

Thresholds – Zero to 100.

Source –

Damos, D. The relation between the Type A behavior pattern, pacing, and subjective workload under single- and dual-task conditions. Human Factors 27(6), 675-680; 1985.

3.2.22 NASA Bipolar Rating Scale

General description - The NASA Bipolar Rating Scale has ten subscales. The titles, endpoints, and descriptions of each scale are presented in Table 23; the scale itself, in Figure 14. If a scale is not relevant to a task, it is given a weight of zero (Hart, Battiste, and Lester, 1984). A weighting procedure is used to enhance intrasubject reliability by 50 percent (Miller and Hart, 1984).

Strengths and limitations - The scale is sensitive to flight difficulty. For example, Bortolussi, Kantowitz, and Hart (1986) reported significant differences in the bipolar ratings between an easy and a difficult flight scenario. Bortolussi, Hart, and Shively (1987) and Kantowitz, Hart, Bortolussi, Shively, and Kantowitz (1984) reported similar results. However, Haworth, Bivens, and Shively (1986) reported that, although the scale discriminated control configurations in a single-pilot configuration, it did not do so in a pilot/copilot configuration. Biferno (1985) reported a correlation between workload and fatigue ratings for a laboratory study. Bortolussi, Kantowitz, and Hart (1986) and Bortolussi, Hart, and Shively (1987) reported that the bipolar scales discriminated two levels of difficulty in a motion-based simulator task. Vidulich and Pandit (1986) reported that the bipolar scales discriminated levels of training in a category search task. Haworth, Bivens, and Shively (1986) reported correlations of +0.79 with Cooper-Harper ratings and +0.67 with SWAT ratings in a helicopter nap-of-the-earth mission. Vidulich and Tsang (1985a, 1985b, 1985c, 1986) reported that the NASA Bipolar Scales were sensitive to task demand, had higher interrater reliability than SWAT, and required less time to complete than SWAT. Vidulich and Bortolussi (1988) reported significant increases in the overall workload rating from cruise to combat phase in simulated helicopter. There was no effect of control configuration.

Data requirements - The number of times a dimension is selected by a subject is used to weight each scale. These weights are then multiplied by the scale score, summed, and divided by the total weight to obtain a workload score. The minimum workload value is zero; the maximum, 100. The scale provides a measure of overall workload but is not sensitive to short-term demands. Further, the activity-type dimension must be carefully explained to pilots before use in flight.

Thresholds - Not stated.

TABLE 23.
NASA Bipolar Rating-Scale Descriptions

Title	Endpoints	Descriptions
Overall Workload	Low, High	The total workload associated with the task considering all sources and components.
Task Difficulty	Low, High	Whether the task was easy, demanding, simple or complex, exacting or forgiving.
Time Pressure	None, Rushed	The amount of pressure you felt due to the rate at which the task elements occurred. Was the task slow and leisurely or rapid and frantic.
Performance	Perfect, Failure	How successful you think you were in doing what we asked you to do and how satisfied you were with what you accomplished.
Mental/Sensory Effort	None, Impossible	The amount of mental and/or perceptual activity that was required (e.g., thinking, deciding, calculating, remembering, looking, searching, etc.).
Physical Effort	None, Impossible	The amount of physical activity that was required (e.g., pushing, pulling, turning, controlling, activating, etc.).
Frustration Level	Fulfilled, Exasperated	How insecure, discouraged, irritated, and annoyed versus secure, gratified, content and complacent you felt.
Stress Level	Relaxed, Tense	How anxious, worried, uptight, and harrassed or calm, tranquil, placid, and relaxed you felt.
Fatigue	Exhausted, Alert	How tired, weary, worn out, and exhausted or fresh, vigorous, and energetic you felt.
Activity Type	Skill based, Rule based, Knowledge based	The degree to which the task required mindless reaction to well-learned routines or required the application of known rules or required problem solving and decision making.

From Lysaght, et al. (1989) p. 91

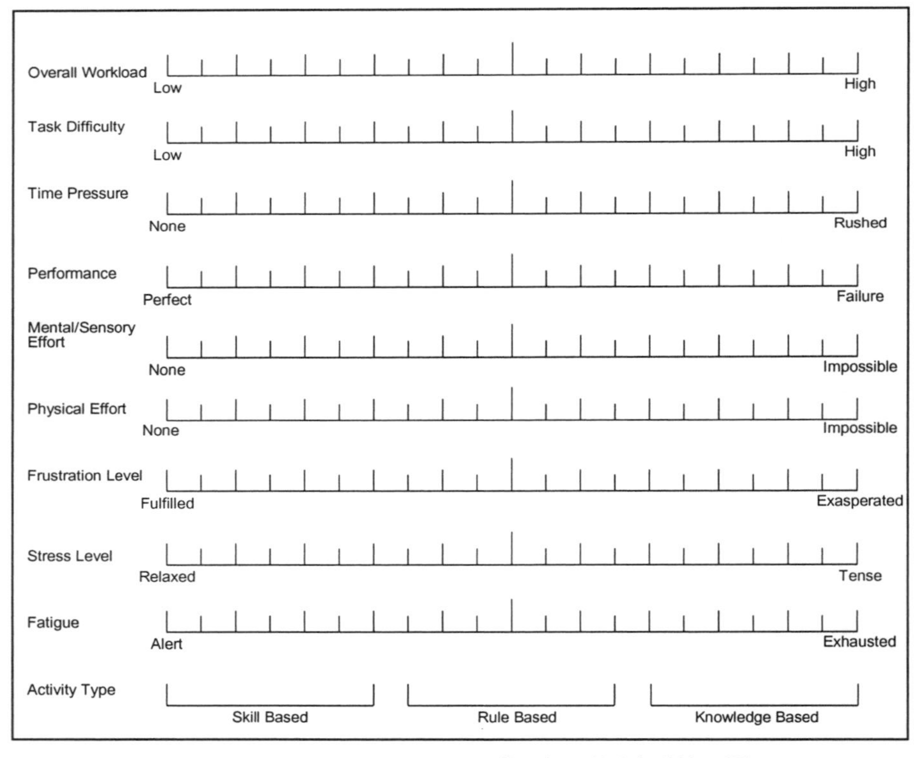

(from Lysaght et al., 1989, p. 92)

FIG. 14. NASA Bipolar Rating Scale

Sources -

Biferno, M.H. Mental workload measurement: Event-related potentials and ratings of workload and fatigue (NASA CR-177354). Washington, DC: NASA; 1985.

Bortolussi, M.R., Hart, S.G., and Shively, R.J. Measuring moment-to-moment pilot workload using synchronous presentations of secondary tasks in a motion-base trainer. In R.S. Jensen (Ed) Proceedings of the 4th Symposium on Aviation Psychology (pp. 651-657). Columbus, OH: Ohio State University; 1987.

Bortolussi, M.R., Kantowitz, B.H., and Hart, S.G. Measuring pilot workload in a motion base trainer: A comparison of four techniques. Applied Ergonomics. 17, 278-283; 1986.

Hart, S.G., Battiste, V., and Lester, P.T. POPCORN: A supervisory control simulation for workload and performance research (NASA-CP-2341). Proceedings of the 20th Annual Conference on Manual Control (pp. 431-453). Washington, DC: NASA; 1984.

Haworth, L.A., Bivens, C.C., and Shively, R.J. An investigation of single-piloted advanced cockpit and control configuration for nap-of-the-earth helicopter combat mission tasks. Proceedings of the 42nd Annual Forum of the American Helicopter Society. 675-671; 1986.

Kantowitz, B.H., Hart, S.G., Bortolussi, M.R., Shively, R.J., and Kantowitz, S.C., Measuring pilot workload in a moving-base simulator: II. Building levels of workload; 1984.

Lysaght, R.J., Hill, S.G., Dick, A.O., Plamondon, B.D., Linton, P.M., Wierwille, W.W., Zaklad, A.L., Bittner, A.C., and Wherry, R.J. Operator workload: comprehensive review and evaluation of operator workload methodologies (Technical Report 851). Alexandria, VA: Army Research Institute for the Behavioral and Social Sciences; June 1989.

Miller, R.C. and Hart, S.G. Assessing the subjective workload of directional orientation tasks (NASA-CP-2341). Proceedings of the 20th Annual Conference on Manual Control (pp. 85-95). Washington, DC: NASA; 1984.

Vidulich, M.A. and Bortolussi, M.R. Speech recognition in advanced rotorcraft: using speech controls to reduce manual control overload. Proceedings of the National Specialists' Meeting Automation Applications for Rotorcraft, 1988.

Vidulich, M.A. and Pandit, P. Training and subjective workload in a category search task. Proceedings of the Human Factors Society 30th Annual Meeting (pp. 1133-1136). Santa Monica, CA: Human Factors Society; 1986.

Vidulich, M.A. and Tsang, P.S. Assessing subjective workload assessment: A comparison of SWAT and the NASA-Bipolar methods. Proceedings of the Human Factors Society 29th Annual Meeting (pp. 71-75). Santa Monica, CA: Human Factors Society; 1985a.

Vidulich, M.A. and Tsang, P.S. Techniques of subjective workload assessment: A comparison of two methodologies. Proceedings of the Third Symposium on Aviation Psychology (pp. 239-246). Columbus, OH: Ohio State University; 1985b.

Vidulich, M.A. and Tsang, P.S. Evaluation of two cognitive abilities tests in a dual-task environment. Proceedings of the 21st Annual Conference on Manual Control (pp. 12.1-12.10). Columbus, OH: Ohio State University; 1985c.

Vidulich, M.A. and Tsang, P.S. Techniques of subjective workload assessment: a comparison of SWAT and NASA-Bipolar methods. Ergonomics. 29(11), 1385-1398; 1986.

3.2.23 NASA Task Load Index

General description - The NASA Task Load Index (TLX) is a multi-dimensional subjective workload rating technique (see Figure 15). In TLX, workload is defined as the "cost incurred by human operators to achieve a specific level of performance." The subjective experience of workload is defined as an integration of weighted subjective responses (emotional, cognitive, and physical) and weighted evaluation of behaviors. The behaviors and subjective responses, in turn, are driven by perceptions of task demand. Task demands can be objectively quantified in terms of magnitude and importance. An experimentally based process of elimination led to the identification of six dimensions for the subjective experience of workload: mental demand, physical demand, temporal demand, perceived performance, effort, and frustration level. The rating-scale definitions are presented in Table 24.

Strengths and limitations - Sixteen investigations were carried out, establishing a database of 3461 entries from 247 subjects (Hart and Staveland, 1987). All dimensions were rated on bipolar scales ranging from 1 to 100, anchored at each end with a single adjective. An overall workload rating was determined from a weighted combination of scores on the six dimensions. The weights were determined from a set of relevance ratings provided by the subjects.

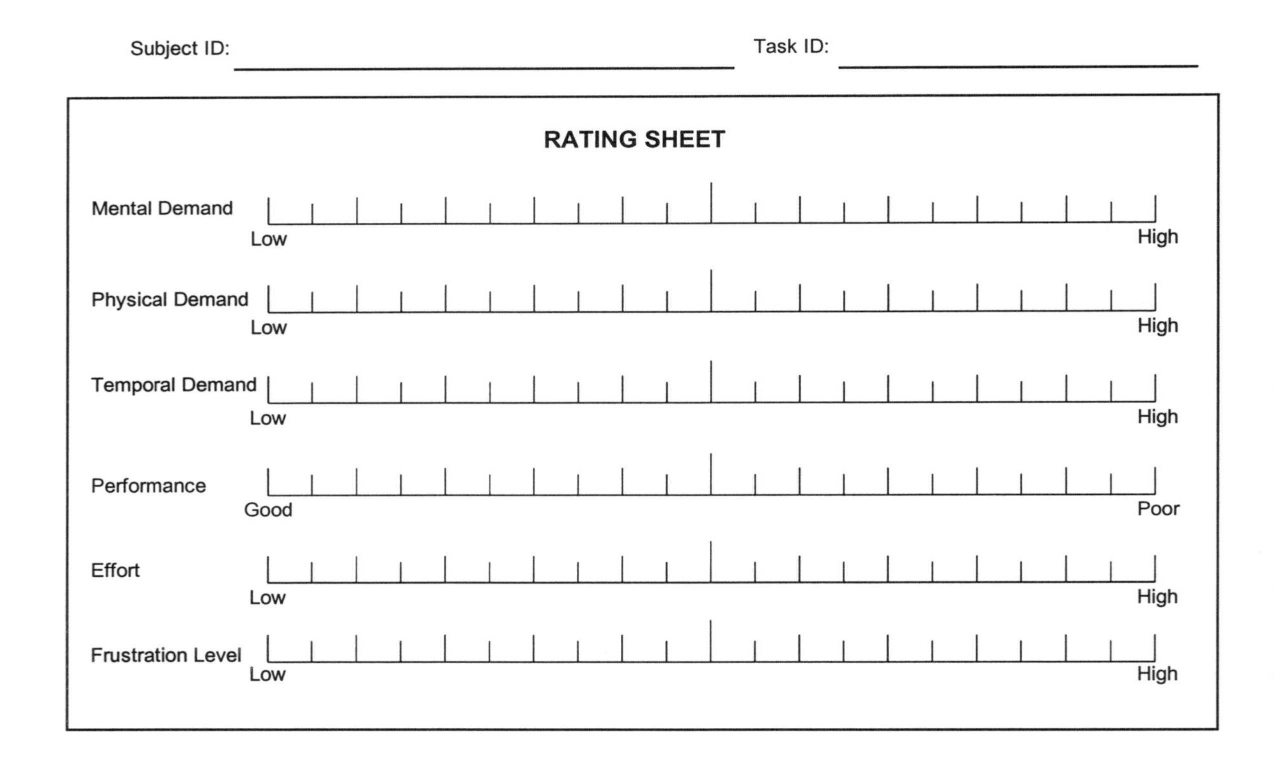
Subject ID: ____________ Task ID: ____________

RATING SHEET

Mental Demand — Low … High

Physical Demand — Low … High

Temporal Demand — Low … High

Performance — Good … Poor

Effort — Low … High

Frustration Level — Low … High

FIG. 15. NASA TLX Rating Sheet

TABLE 24.
NASA TLX Rating-Scale Descriptions

Title	Endpoints	Descriptions
Mental Demand	Low, High	How much mental and perceptual activity was required (e.g., thinking, deciding, calculating, remembering, looking, searching, etc.)? Was the task easy or demanding, simple or complex, exacting or forgiving?
Physical Demand	Low, High	How much physical activity was required (e.g., pushing, pulling, turning, controlling, activating, etc.)? Was the task easy or demanding, slow or brisk, slack or strenuous, restful or laborious?
Temporal Demand	Low, High	How much time pressure did you feel due to the rate or pace at which the tasks or task elements occurred? Was the pace slow and leisurely or rapid and frantic?
Performance	Good, Poor	How successful do you think you were in accomplishing the goals of the task set by the experimenter (or yourself)? How satisfied were you with your performance in accomplishing these goals?
Effort	Low, High	How hard did you have to work (mentally and physically) to accomplish your level of performance?
Frustration Level	Low, High	How insecure, discouraged, irritated, stressed, and annoyed versus secure, gratified, content, relaxed and complacent did you feel during the task? (NASA Task Load Index, p. 13)

Hart and Staveland (1987) concluded that the TLX provides a sensitive indicator of overall workload as it differed among tasks of various cognitive and physical demands. They also stated that the weights and magnitudes determined for each TLX dimension provide important diagnostic information about the sources of loading within a task. They reported that the six TLX ratings took less than a minute to acquire and suggested the scale would be useful in operational environments.

Battiste and Bortolussi (1988) reported significant workload effects as well as a test-retest correlation of +0.769. Corwin, Sandry-Garza, Biferno, Boucek, Logan, Jonsson, and Metalis (1989) reported that NASA TLX was a valid and reliable measure of workload.

NASA TLX has been used extensively in the flight environment. Bittner, Byers, Hill, Zaklad, and Christ (1989), Byers, Bittner, Hill, Zaklad, and Christ (1988), Hill, Byers, Zaklad, and Christ (1989), Hill, Zaklad, Bittner, Byers, and Christ (1988), and Shively, Battiste, Matsumoto, Pepitone, Bortolussi, and Hart (1987), based on in-flight data, stated that TLX ratings significantly discriminated flight segments.

Nataupsky and Abbott (1987) successfully applied NASA TLX to a multi-task environment. Vidulich and Bortolussi (1988) replicated the significant flight-segment effect but reported no significant differences in TLX ratings between control configurations or between combat countermeasure conditions. In a later study, Tsang and Johnson (1989) reported reliable increases in NASA TLX ratings when target-acquisition and engine-failure tasks were added to the primary flight task. Vidulich and Bortolussi (1988) reported significant increases in NASA TLX ratings from the cruise to the combat phase during a simulated helicopter mission. There was no affect of control configuration, however.

Nygren (1991) reported that NASA TLX is a measure of general workload experienced by aircrews. Selcon, Taylor, and Koritsas (1991) concluded from pilot ratings of an air combat flight simulation that NASA TLX was sensitive to difficulty but not the pilot experience. Hancock, Williams, Manning, and Miyake (1995) reported the NASA TLX score was highly correlated with difficulty of a simulated flight task.

Hendy, Hamilton, and Landry (1993) examined undimensional and multidimensional measures of workload in a series of four experiments (low-level helicopter operations, peripheral version display evaluation, flight simulator fidelity, and aircraft landing task). They concluded that if an overall measure of workload is required, then a univariate measure is as sensitive as an estimate derived from multivariate data. If a univariate measure is not available then a simple unweighted additive method can be used to combine ratings into an overall workload estimate.

Byers, Bittner, and Hill (1989) suggested using raw TLX scores. Moroney, Biers, Eggemeier, and Mitchell (1992) reported that the pre-rating weighting scheme is unnecessary since the correlation between weighted and unweighted scores was +0.94. Further, delays of 15 minutes did not affect the workload ratings; delays of 48 hours, however, did. After 48 hours, ratings no longer discriminate workload conditions. Moroney, Biers, and Eggemeier (1995) concluded from a review of relevant studies that 15-minute delays do not affect NASA TLX.

TLX has been applied to other environments. Hill, Iavecchia, Byers, Bittner, Zaklad, and Christ (1992) reported that the NASA TLX was sensitive to different levels of workload and high in user acceptance. Their subjects were Army operators. Jordan and Johnson (1993) concluded from an on-road evaluation of a car stereo that TLX was a useful measure of mental workload.

Hancock and Caird (1993) reported a significant increase in the overall workload rating scale of the NASA TLX as shrink rate of a target decreased. The highest ratings were on paths with four steps rather than 2, 8, or 16 steps from cursor to target. NASA TLX scores significantly increased as ambient noise increased (Becker, Warm, Dember, and Hancock, 1995).

Harris, Hancock, Arthur, and Caird (1995) reported significantly higher ratings on five (mental demand, temporal demand, effort, frustration, and physical demand) of the six NASA-TLX scales for manual than automatic tracking.

Vidulich and Tsang (1985) compared the SWAT and TLX. They stated that the collection of ratings is simpler with SWAT. However, the SWAT card sort is more tedious and time consuming. Battiste and Bortolussi (1988) reported no significant correlation between SWAT and NASA TLX in a simulated B-727 flight. Hancock (1996) stated that NASA TLX and SWAT "were essentially equivalent in terms of their sensitivity to task manipulations." The task was tracking. Tsang and Johnson (1987) reported good correlations between NASA TLX and a one dimensional workload scale. Vidulich and Tsang (1987) replicated the Tsang and Johnson finding as well as reported a good correlation between NASA TLX and the Analytical Hierarchy Process.

Vidulich and Pandit (1987) reported only three significant correlations between NASA TLX and seven personality tests (Jenkins Activity Survey, Rotter's Locus of Control, Cognitive Failures Questionnaire, Cognitive Interference Questionnaire, Thought Occurrence Questionnaire, California Q-Sort, and the Myers-Briggs Type Indicator): the speed scale of the Jenkins Activity Survey and the physical demand scale of the NASA TLX ($r = -0.23$), Locus of control and physical demand ($r = +0.21$), and finally locus of control and effort ($r = +0.23$).

Data requirements - Use of the TLX requires two steps. First, subjects rate each task performed on each of the six subscales. Hart suggests that subjects should practice using the rating scales in a training session. Second, subjects must perform 15 pair-wise comparisons of six workload scales. The number of times each scale is rated as contributing more to the workload of a task is used as the weight for that scale. Separate weights should be derived for diverse tasks; the same weights can be used for similar tasks. Note that a set of IBM PC compatible programs has been written to gather ratings and weights and to compute the weighted workload scores. The programs are available from the Human Factors Division at NASA Ames Research Center, Moffett Field, CA.

Thresholds - Knapp and Hall (1990) used NASA TLX to evaluate a highly automated communication system. Using 40 as a high workload threshold, the system was judged to impose high workload and difficult cognitive effort on operators.

Sources -

Battiste, V. and Bortolussi, M.R. Transport pilot workload: a comparison of two objective techniques. Proceedings of the Human Factors Society 32nd Annual Meeting. 150-154; 1988.

Becker, A.B., Warm, J.S., Dember, W.N., and Hancock, P.A. Effects of jet engine noise and performance feedback on perceived workload in a monitoring task. International Journal of Aviation Psychology. 5(1), 49-62, 1995.

Bittner, A.C., Byers, J.C., Hill, S.G., Zaklad, A.L., and Christ, R.E. Generic workload ratings of a mobile air defense system. Proceedings of the Human Factors Society 33rd Annual Meeting (pp. 1476-1480). Santa Monica, CA: Human Factors Society; 1989.

Byers, J.C., Bittner, A.C., and Hill, S.G. Traditional and raw Task Load Index (TLX) correlations: are paired comparisons necessary? In Advances in industrial erogonomics and safety. London: Taylor and Frances; 1989.

Byers, J.C., Bittner, A.C., Hill, S.G., Zaklad, A.L., and Christ, R.E. Workload assessment of a remotely piloted vehicle (RPV) system. Proceedings of the Human Factors Society 32nd Annual Meeting (pp. 1145-1149). Santa Monica, CA: Human Factors Society; 1988.

Corwin, W.H., Sandry-Garza D.L., Biferno, M.H., Boucek, G.P., Logan, A.L., Jonsson, J.E., and Metalis, S.A. Assessment of crew workload measurement methods, techniques, and procedures. Volume I-Process, methods, and results (WRDC-TR-89-7006). Wright-Patterson Air Force Base, OH; 1989.

Hancock, P.A. Effects of control order, augmented feedback, input device, and practice on tracking performance and perceived workload. Ergonomics. 39(9), 1146-1162, 1996.

Hancock, P.A. and Caird, J.K. Experimental evaluation of a model of mental workload. Human Factors. 35(3), 413-419; 1993.

Hancock, P.A., William G., Manning, C.M., and Miyake, S. Influence of task demand characteristics on workload and performance. International Journal of Aviation Psychology. 5(1), 63-86, 1995.

Harris, W.C., Hancock, P.A., Arthur, E.J., and Caird, J.K. Performance, workload, and fatigue changes associated with automation. International Journal of Aviation Psychology. 5(2), 169-185; 1995.

Hart, S.G. and Staveland, L.E. Development of NASA-TLX (Task Load Index): Results of empirical and theoretical research. In P.A. Hancock and N. Meshkati (Eds) Human mental workload. Amsterdam: Elsevier; 1987.

Hendy, K.C., Hamilton, K.M., and Landry, L.N. Measuring subjective workload: when is one scale better than many? Human Factors. 35(4), 579-601; 1993.

Hill, S.G., Byers, J.C., Zaklad, A.L., and Christ, R.E. Subjective workload assessment during 48 continuous hours of LOS-F-H operations. Proceedings of the Human Factors Society 33rd Annual Meeting (pp. 1129-1133). Santa Monica, CA: Human Factors Society; 1989.

Hill, S.G., Iavecchia, H.P., Byers, J.C., Bittner, A.C., Zaklad, A.L., and Christ, R.E. Comparison of four subjective workload rating scales. Human Factors. 34:429-439; 1992.

Hill, S.G., Zaklad, A.L., Bittner, A.C., Byers, J.C., and Christ, R.E. Workload assessment of a mobile air defense system. Proceedings of the Human Factors Society 32nd Annual Meeting (pp. 1068-1072). Santa Monica, CA: Human Factors Society; 1988.

Jordan, P.W. and Johnson, G.L. Exploring mental workload via TLX: the case of operating a car stereo whilst driving. In A.G. Gale, I.D. Brown, C.M. Haslegrave, H.W. Kruysse, and S.P. Taylor (Eds.). Vision in vehicles - IV. Amsterdam: North-Holland; 1993.

Knapp, B.G. and Hall, B.J. High performance concerns for the TRACKWOLF system (ARI Research Note 91-14). Alexandria, VA; 1990.

Moroney, W.F., Biers, D.W., and Eggemeier, F.T. Some measurement and methodological considerations in the application of subjective workload measurement techniques. International Journal of Aviation Psychology. 5(1), 87-106, 1995.

Moroney, W.E., Biers, D.W., Eggemeier, F.T., and Mitchell, J.A. A comparison of two scoring procedures with the NASA Task Load Index in a simulated flight tasks. NAECON Proceedings (pp. 734-740). Dayton, OH; 1992.

Nataupsky, M. and Abbott, T.S. Comparison of workload measures on computer-generated primary flight displays. Proceedings of the Human Factors Society 31st Annual Meeting (pp. 548-552). Santa Monica, CA: Human Factors Society; 1987.

Nygren, T.E. Psychometric properties of subjective workload measurement techniques: Implications for their use in the assessment of perceived mental workload. Human Factors. 33 (1), 17-33; 1991.

Selcon, S.J., Taylor, R.M., and Koritsas, E. Workload or situational awareness?: TLX vs. SART for aerospace systems design evaluation. Proceedings of the Human Factors Society 35th Annual Meeting. 62-66, 1991.

Shively, R.J., Battiste, V., Matsumoto, J.H., Pepitone, D.D., Bortolussi, M.R., and Hart, S.G. Inflight evaluation of pilot workload measures for rotorcraft research. In R.S. Jensen Proceedings of the 4th Symposium on Aviation Psychology (pp. 637-643). Columbus, OH: Ohio State University; 1987.

Tsang, P.S. and Johnson, W. Automation: Changes in cognitive demands and mental workload. Proceedings of the Fourth Symposium on Aviation Psychology. Columbus, OH: Ohio State University; 1987.

Tsang, P.S. and Johnson, W.W. Cognitive demands in automation. Aviation, Space, and Environmental Medicine. 60, 130-135; 1989.

Vidulich, M.A. and Bortolussi, M.R. Control configuration study. Proceedings of the American Helicopter Society National Specialist's Meeting: Automation Application for Rotorcraft; 1988.

Vidulich, M.A. and Bortolussi, M.R. Speech recognition in advanced rotorcraft: Using speech controls to reduce manual control overload. Proceedings of the National Specialists' Meeting Automation Applications for Rotorcraft, 1988.

Vidulich, M.A. and Pandit, P. Individual differences and subjective workload assessment: Comparing pilots to nonpilots. Proceedings of the International Symposium on Aviation Psychology. 630-636, 1987.

Vidulich, M.A. and Tsang, P.S. Assessing subjective workload assessment: A comparison of SWAT and the NASA-bipolar methods. Proceedings of the Human Factors Society 29th Annual Meeting (pp. 71-75). Santa Monica, CA: Human Factors Society; 1985.

Vidulich, M.A. and Tsang, P.S. Absolute magnitude estimation and relative judgment approaches to subjective workload assessment. Proceedings of the Human Factors Society 31st Annual Meeting (pp. 1057-1061). Santa Monica, CA: Human Factors Society; 1987.

3.2.24 Overall Workload Scale

General description - The Overall Workload (OW) Scale is a bipolar scale ("low" on the left; "high" on the right) requiring subjects to provide a single workload rating on a horizontal line divided into 20 equal intervals.

Strengths and limitations - The OW scale is easy to use but is less valid and reliable than NASA Task Load Index (TLX) or Analytical Hierarchy Process (AHP) ratings (Vidulich and Tsang, 1987). Hill, Iavecchia, Byers, Bittner, Zaklad, and Christ (1992) reported that OW was consistently more sensitive to workload and had greater operator acceptance than the Modified Cooper-Harper rating scale or the Subjective Workload Assessment Technique (SWAT). Harris, Hill, Lysaght, and Christ (1992) reported that the overall workload scale has been sensitive across tasks, systems, and environments. The scale can be used retrospectively or prospectively (Eggleston and Quinn, 1984). It has been used in assessing workload in mobile air defense missile system (Hill, Zaklad, Bittner, Byers, and Christ 1988), remotely piloted vehicle systems (Byers, Bittner, Hill, Zaklad, and Christ, 1988), helicopter simulators (Iavecchia, Linton, and Byers, 1989); and laboratories (Harris, Hancock, Arthur, and Caird, 1995).

Data requirements - Not stated.

Thresholds - Not stated.

Sources -

Byers, J.C., Bittner, A.C., Hill, S.G., Zaklad, A.L., and Christ, R.E. Workload assessment of a remotely piloted vehicle (RPV) system. Proceedings of the Human Factors Society 32nd Annual Meeting (pp. 1145-1149). Santa Monica, CA: Human Factors Society; 1988.

Eggleston, R.G. and Quinn, T.J. A preliminary evaluation of a projective workload assessment procedure. Proceedings of the Human Factors Society 28th Annual Meeting (pp. 695-699). Santa Monica, CA: Human Factors Society; 1984.

Harris, W.C., Hancock, P.A., Arthur, E.J., and Caird, J.K. Performance, workload, and fatigue changes associated with automation. International Journal of Aviation Psychology. 5(2), 169-185; 1995.

Harris, R.M., Hill, S.G., Lysaght, R.J., and Christ, R.E. Handbook for operating the OWLKNEST technology (ARI Research note 92-49). Alexandria, VA: United States Army Research Institute for the Behavioral and Social Sciences; 1992.

Hill, S.G., Iavecchia, H.P., Byers, J.C., Bittner, A.C., Zaklad, A.L., and Christ, R.E. Comparison of four subjective workload rating scales. Human Factors. 34: 429-439; 1992.

Hill, S.G., Zaklad, A.L., Bittner, A.C., Byers, J.C., and Christ, R.E. Workload assessment of a mobile air defense missile system. Proceedings of the Human Factors Society 32nd Annual Meeting (pp. 1068-1072). Santa Monica, CA: Human Factors Society; 1988.

Iavecchia, H.P., Linton, P.M., and Byers, J.C. Operator workload in the UH-60A Black Hawk crew results vs. TAWL model predictions. Proceedings of the Human Factors Society 33rd Annual Meeting (pp. 1481-1481). Santa Monica, CA: Human Factors Society; 1989.

Vidulich, M.A. and Tsang, P.S. Absolute magnitude estimation and relative judgement approaches to subjective workload assessment. Proceedings of the Human Factors Society 31st Annual Meeting. 1057-1061; 1987.

3.2.25 Pilot Objective/Subjective Workload Assessment Technique

General description - The Pilot Objective/Subjective Workload Assessment Technique (POSWAT) is a ten-point subjective scale developed at the Federal Aviation Administration's Technical Center (Stein, 1984). The scale is a modified Cooper-Harper scale but does not include the binary decision tree that is characteristic of the Cooper-Harper scale. It does, however, divide workload into five categories: low, minimal, moderate, considerable, and excessive. Like the Cooper-Harper, the lowest three levels (1 through 3) are grouped into a low category. A similar scale, the Air Traffic Workload Input Technique (ATWIT), has been developed for air traffic controllers (Porterfield, 1997).

Strengths and limitations - The immediate predecessor of POSWAT was the Workload Rating System. It consisted of a workload entry device with an array of 10 pushbuttons. Each pushbutton corresponded to a rating from 1 (very easy) to 10 (very hard). The scale was sensitive to changes in flight control stability (Rehmann, Stein, and Rosenberg, 1983).

Stein (1984) reported that POSWAT ratings significantly differentiated experienced and novice pilots and high (initial and final approach) and low (en route) flight segments. There was also a significant learning effect: workload ratings were significantly higher on the first than on the second flight. Although the POSWAT scale was sensitive to manipulations of pilot experience level for flights in a light aircraft and in a simulator, the scale was cumbersome. Seven dimensions (workload, communications, control inputs, planning, "deviations," error, and pilot complement) are combined on one scale. Further, the number of ranks on the ordinal scale are confusing since there are both five and ten levels.

Rehman, Stein, and Rosenberg (1983) obtained POSWAT ratings once per minute. These investigators found that pilots reliably reported workload differences in a tracking task on a simple ten-point non-adjectival scale. Therefore, the cumbersome structure of the POSWAT scale may not be necessary.

Data requirements - Stein (1984) suggested not analyzing POSWAT ratings for short flight segments if the ratings are given at one-minute intervals.

Thresholds - Not stated.

Sources -

Porterfield, D.H. Evaluating controller communication time as a measure of workload. The International Journal of Aviation Psychology. 7(2): 171-182; 1997.

Rehman, J.T., Stein, E.S., and Rosenberg, B.L. Subjective pilot workload assessment. Human Factors. 25(3): 297-307; 1983.

Stein, E.S. The measurement of pilot performance: A master-journeyman approach (DOT/FAA/CT-83/15). Atlantic City, NJ: Federal Aviation Administration Technical Center; May 1984.

3.2.26 Pilot Subjective Evaluation

General description - The Pilot Subjective Evaluation (PSE) workload scale (see Figure 16) was developed by Boeing for use in the certification of the Boeing 767 aircraft. The scale is accompanied by a questionnaire. Both the scale and the questionnaire are completed with reference to an existing aircraft selected by the pilot.

Strengths and limitations - Fadden (1982) and Ruggerio and Fadden (1987) stated that the ratings of workload greater than the reference aircraft were useful in identifying aircraft design deficiencies.

Data requirements - Each subject must complete both the PSE scale and the questionnaire.

Thresholds - 1, minimum workload; 7, maximum workload.

Sources -

Fadden, D. Boeing Model 767 flight deck workload assessment methodology. Paper presented at the SAE Guidance and Control System Meeting, Williamsburg, VA; 1982.

Lysaght, R.J., Hill, S.G., Dick, A.O., Plamondon, B.D., Linton, P.M., Wierwille, W.W., Zaklad, A.L., Bittner, A.C., and Wherry, R.J. Operator workload: comprehensive review and evaluation of operator workload methodologies (Technical Report 851). Alexandria, VA: Army Research Institute for the Behavioral and Social Sciences; June 1989.

Ruggerio, F. and Fadden, D. Pilot subjective evaluation of workload during a flight test certification programme. In A.H. Roscoe (Ed.) The practical assessment of pilot workload. AGARD-ograph 282 (pp. 32-36). Neuilly-sur-Seine, France: AGARD; 1987.

3.2.27 Profile of Mood States

General description - The shortened version of the Profile of Mood States (POMS) scale (Shachem, 1983) provides measures of self-rated tension, depression, anger, vigor, fatigue, and confusion.

PILOT SUBJECTIVE EVALUATION SCALE

Mental Effort
More Less

Physical Difficulty
More Less

Time Required
More Less

Understanding of Horizontal Position
More Less

Navigation

FMS Operation and Monitoring

Engine Airplane Systems Operating and Monitoring

Manual Flight Path Control

Communications

Command Decisions

Collision Avoidance

(Complete For Selected Tasks)

Time Available
More Less

Usefulness of Information
More Less

(from Lysaght, et al., 1989, p. 107)

FIG. 16. Pilot Subjective Evaluation Scale

Strengths and limitations - Reliability and validation testing of the POMS has been extensive. For example, McNair and Lorr (1964) reported test/retest reliabilities of 0.61 to 0.69 for the six factors. Reviews of the sensitivity and reliability of the POMS have been favorable (Norcross, Guadagnoli, and Prochaska, 1984). Constantini, Braun, Davis, and Iervolino (1971) reported significant positive correlations between POMS and the Psychological Screening Inventory, thus yielding consensual validation. Pollock, Cho, Reker, and Volavka (1979) correlated POMS scales and physiological measures from eight healthy males. The tension and depression scores were significantly correlated with heart rate (+0.75 and +0.76, respectively) and diastolic blood pressure (+0.71 and +0.72, respectively). Heart rate was also significantly correlated with the anger score (+0.70).

The POMS has been used extensively in psychotherapy research (e.g., Haskell, Pugatch, and McNair, 1969; Lorr, McNair, Weinstein, Michaux, and Raskin, 1961; McNair, Goldstein, Lorr, Cibelli, and Roth, 1965; Pugatch, Haskell, and McNair, 1969) and drug research (e.g., Mirin, Shapiro, Meyer, Pillard, and Fisher, 1971; Nathan, Titler, Lowenstein, Solomon, and Rossi, 1970; Nathan, Zare, Ferneau, and Lowenstein, 1970; Pillard and Fisher, 1970).

Storm and Parke (1987) used the POMS to assess the mood effects of a sleep-inducing drug (temazepam) for EF-111 aircrews. As hoped, there were no significant drug effects on any of the six subscales. Gawron, et al. (1988) asked subjects to complete the POMS after a 1.75-hour flight. There were no significant crew position effects on rated vigor or fatigue. There was a significant order effect on fatigue, however. Subjects who had been pilots first had higher ratings (2.7) than subjects who had been copilots first (1.3).

Harris, Hancock, Arthur, and Caird (1995) did not find a significant difference in the fatigue rating between a manual and an automatic tracking group.

Data requirements - The POMS takes about 10 minutes to complete and requires a stiff writing surface. The POMS is available from the Educational and Industrial Testing Service, San Diego, CA.

Thresholds - Not stated.

Sources -

Costantini, A.F., Braun, J.R., Davis, J.E., and Iervolino, A. The life change inventory: a device for quantifying psychological magnitude of changes experienced by college students. Psychological Reports, 34 (3, Pt. 1), 991-1000; June 1971.

Gawron, V.J., Schiflett, S., Miller, J., Ball, J., Slater, T., Parker, F., Lloyd, M., Travale, D., and Spicuzza, R.J. The effect of pyridostigmine bromide on inflight aircrew performance (USAFSAM-TR-87-24). Brooks Air Force Base, TX: School of Aerospace Medicine; January 1988.

Harris, W.C., Hancock, P.A., Arthur, E.J., and Caird, J.K. Performance, workload, and fatigue changes associated with automation. International Journal of Aviation Psychology. 5(2), 169-185; 1995.

Haskell, D.H., Pugatch, D. and McNair, D.M. Time-limited psychotherapy for whom? Archives of General Psychiatry. 21, 546-552; 1969.

Lorr, M., McNair, D.M., Weinstein, G.J., Michaux, W.W., and Raskin, A. Meprobromate and chlorpromazin in psychotherapy. Archives of General Psychiatry. 4, 381-389; 1961.

McNair, D.M., Goldstein, A.P., Lorr, M., Cibelli, L.A., and Roth, I. Some effects of chlordiazepoxide and meprobromate with psychiatric outpatients. Psychopharmacologia. 7, 256-265; 1965.

McNair, D.M., and Lorr, M. An analysis of mood in neurotics. Journal of Abnormal Psychology. 69, 620-627; 1964.

Mirin, S.M., Shapiro, L.M., Meyer, R.E., Pillard, R.C. and Fisher, S. Casual versus heavy use of marijuana: A redefinition of the marijuana problem. American Journal of Psychiatry. 172, 1134-1140; 1971.

Nathan, P.F., Titler, N.A., Lowenstein, L.M., Solomon, P., and Rossi, A.M. Behavioral analyses of chronic alcoholism: Interaction of alcohol and human contact. Archives of General Psychiatry. 22, 419-430; 1970.

Nathan, P.F., Zare, N.C., Ferneau, E.W. and Lowenstein, L.M. Effects of congener differences in alcohol beverages on the behavior of alcoholics. Quarterly Journal on Studies of Alcohol. Supplement No. 5. 87-100; 1970.

Norcross, J.C., Guadagnoli, E., and Prochaska, J.O. Factor structure of the profile of mood states (POMS): Two partial replications. Journal of Clinical Psychology. 40, 1270-1277; 1984.

Pillard, R.C. and Fisher, S. Aspects of anxiety in dental clinic patients. Journal of the American Dental Association. 80, 1331-1334; 1970.

Pollock, V., Cho, D.W., Reker, D., and Volavka, J. Profile of mood states: The factors and their correlates. Journal of Nervous Mental Disorders. 167, 612-614; 1979.

Pugatch, D., Haskell, D.H., and McNair, D.M. Predictors and patterns of change associated with the course of time limited psychotherapy (Mimeo Report); 1969.

Shachem, A. A shortened version of the profile of mood states. Journal of Personality Assessment. 47, 305-306; 1983.

Storm, W.F., and Parke, R.C. FB-111A aircrew use of temazepam during surge operations. Proceedings of NATO Advisory Group for Aerospace Research and Development (AGARD) Biochemical Enhancement of Performance Conference (Paper No. 415, pp. 12-1 to 12-12). Neuilly-sur-Seine, France: AGARD; 1987.

3.2.28 Sequential Judgment Scale

General description – Pitrella and Kappler (1988) developed the Sequential Judgment Scale to measure the difficulty of driver vehicle handling. It was designed to meet the following rating scale guidelines: "1) use continuous instead of category scale formats, 2) use both verbal descriptors and numbers at scale points, 3) use descriptors at all major scale markings, 4) use horizontal rather than vertical scale formats, 5) either use extreme or no descriptors at end points, 6) use short, precise, and value-unloaded descriptors, 8) select and use equidistant descriptors, 9) use psychologically-scaled descriptors, 10) use positive numbers only, 11) have desirable qualities increase to the right, 12) use descriptors free of evaluation demands and biases, 13) use 11 or more scale points as available descriptors permit, and 14) minimize rater workload with suitable aids" (Pfendler, Pitrella, and Wiegand, 1994, p. 28). The scale has interval scale properties. It exists in both 11- and 15-point versions in German, Dutch, and English. The 15-point English version is presented in Figure 17.

Strength and limitations – Kappler, Pitrella, and Godthelp (1988) reported that the Sequential Judgment Scale ratings varied significantly between loaded and unloaded trucks as well as between different models of trucks. Kappler and Godthelp (1989) reported significantly more difficult in vehicle handling as tire pressure and lane width decreased. The subjects drove on a closed-loop straight lane.

Pitrella (1988) reported that the scale significantly discriminated ten difficulty levels in a tracking task. Difficulty was manipulated by varying the amplitude and frequency of the forcing function. Pfendler (1993) reported higher validity estimates for the Sequential Judgment Scale (+0.72) than the German-version of the NASA TLX (+0.708) in a color detection task.

Reliabilities have also been high (+0.92 to +0.99, Kappler, Pitrella, and Godthelp, 1988; +0.87 to +0.99, Pitrella, 1989; +0.87, Pfendler, 1993). Since the scale is an interval scale, parametric statistics can be used to analyze the data.

The scale has two disadvantages: "1) if only overall workload is measured, rating results will have low diagnosticity and 2) information on validity of the scale is restricted to psychomotor and perceptual tasks" (Pfendler, Pitrella, and Wiegand, 1994, p. 30).

Data requirements – Subjects mark the scale in pen or pencil. The experimenter then measures the distance from the right end of the scale. This measure is converted to a percentage of the complete scale.

Thresholds – 0 to 100%.

Sources –

Kappler, W.D. and Godthelp, H. Design and use of the two-level Sequential Judgment Rating Scale in the identification of vehicle handling criteria: I. Instrumented car experiments on straight lane driving. Wachtberg: Forschungsinstitut fur Anthropotechnik, FAT Report Number 79, 1989.

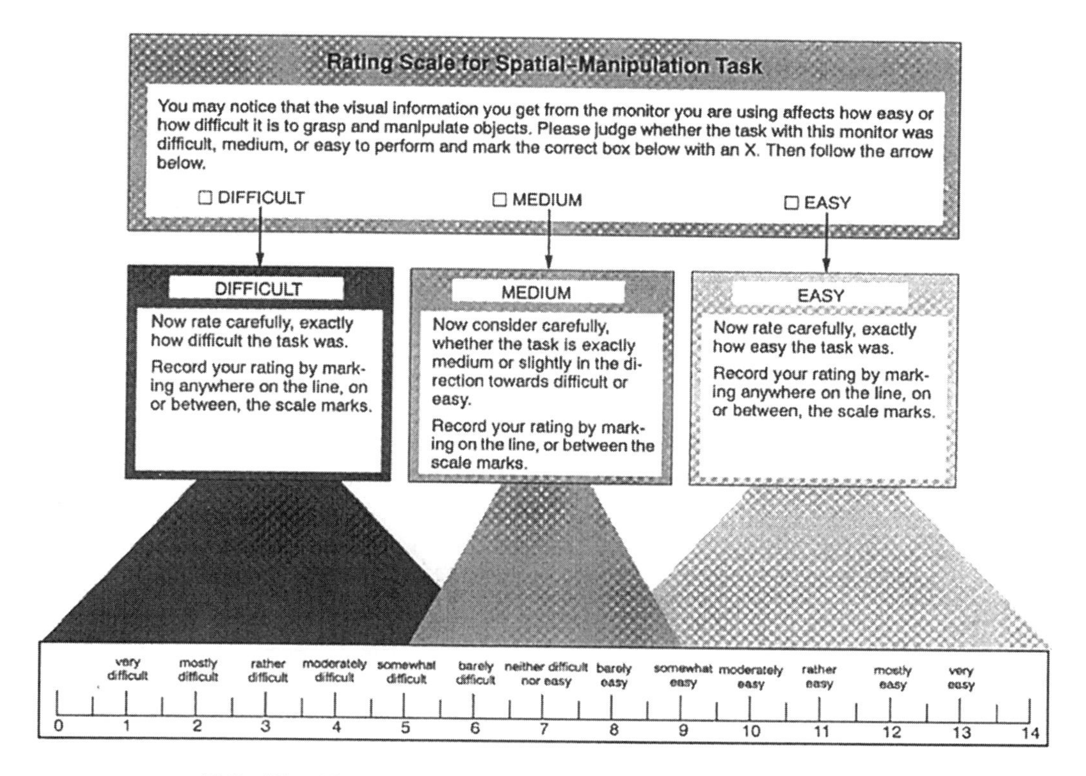

FIG. 17. 15-point Form of the Sequential Judgment Scale (Pfender, Pitrella, and Wiegand, 1994, p. 31).

Kappler, W.D., Pitrella, F.D., and Godthelp, H. Psychometric and performance measurement of light weight truck handling qualities. Wachtberg: Forschungsinstitut fur Anthropotechnik, FAT Report Number 77, 1988.

Pfendler, C. Vergleich der Zwei-Ebenen Intensitats-Skala und des NASA Task Load Index bei de Beanspruchungsbewertung wahrend ternvorgangen. Z. Arb. Wise 47 (19 NF) 1993/1, 26-33.

Pfendler, C., Pitrella, F.D., and Wiegand, D. Workload measurement in human engineering test and evaluation. Forschungsinstitut fur Anthropotechnik, Bericht Number 109, July 1994.

Pitrella, F.D. A cognitive model of the internal rating process. Wachtberg: Forschungsinstitut fur Anthropotechnik, FAT Report Number 82, 1988.

Pitrella, F.D. and Kappler, W.D. Identification and evaluation of scale design principles in the development of the sequential judgment, extended range scale. Wachtberg: Forschungsinstitut fur Anthropotechnik, FAT Report Number 80, 1988.

3.2.29 Subjective Workload Assessment Technique

General description - The Subjective Workload Assessment Technique (SWAT) combines ratings of three different scales (see Table 25) to produce an interval scale of mental workload. These scales are: (1) time load, which reflects the amount of spare time available in planning, executing, and monitoring a task; (2) mental effort load, which assesses how much conscious mental effort and planning are required to perform a task; and (3) psychological stress load, which measures the amounts of risk, confusion, frustration, and anxiety associated with task performance. A more complete description is given in Reid and Nygren (1988). A description of the initial conjoint measurement model for SWAT is described in Nygren (1982, 1983).

TABLE 25.
SWAT Scales

Time Load
1. Often have spare time. Interruptions or overlap among activities occur infrequently or not at all.
2. Occasionally have spare time. Interruptions or overlap among activities occur frequently.
3. Almost never have spare time. Interruptions or overlap among activities are frequent or occur all the time.

Mental Effort Load
1. Very little conscious mental effort or concentration required. Activity is almost automatic, requiring little or no attention.
2. Moderate conscious mental effort or concentration required. Complexity of activity is moderatelyhigh due to uncertainty, unpredictability, or unfamiliarity. Considerable attention required.
3. Extensive mental effort and concentration are necessary. Very complex activity requiring total attention.

Psychological Stress Load
1. Little confusion, risk, frustration, or anxiety exists and can be easily accommodated.
2. Moderate stress due to confusion, frustration, or anxiety noticeably adds to workload. Significant compensation is required to maintain adequate performance.
3. High to very intense stress due to confusion, frustration, or anxiety. High to extreme determination and self-control required. (Potter and Bressler, 1989, pp. 12-14).

Strengths and limitations - SWAT has been found to be a valid (Albery, Repperger, Reid, Goodyear, and Roe, 1987; Haworth, Bivens, and Shively, 1986; Masline, 1986; Reid, Shingledecker, and Eggemeier, 1981; Reid, Shingledecker, Nygren, and Eggemeier, 1981; Vidulich and Tsang, 1987; Warr, Colle, and Reid, M.G, 1986), sensitive (Eggemeier, Crabtree, Zingg, Reid, and Shingledecker, 1982), reliable (Corwin, Sandry-Garza, Biferno, Boucek, Logan, Jonsson, and Metalis, 1989; Gidcomb, 1985), and relatively unobtrusive (Crabtree, Bateman, and Acton, 1984; Courtright and Kuperman, 1984; Eggemeier, 1988) measure of workload. Further, SWAT ratings are not affected by delays of up to 30 minutes (Eggemeier, Crabtree, and LaPoint, 1983), nor by intervening tasks of all but difficult tasks (Eggemeier, Melville, and Crabtree, 1984; Lutmer and Eggemeier, 1990). Moroney, Biers, and Eggemeier (1995) concur. Also, Eggleston (1984) found a significant correlation between projected SWAT ratings made during system concept evaluation and those made during ground-based simulation of the same system.

Warr (1986) reported that SWAT ratings were less variable than modified Cooper-Harper ratings. Kilmer, et al. (1988) reported that SWAT was more sensitive to changes in difficulty of a tracking task than the modified Cooper-Harper Rating Scale was. Finally, Nygren (1991) stated that SWAT provides a good cognitive model of workload, sensitive to individual differences.

SWAT has been used in diverse environments, for example, test aircraft (Papa and Stoliker, 1988), a high-G centrifuge (Albery, Ward, and Gill, 1985), command, control, and communications centers (Crabtree, Bateman, and Acton, 1984), nuclear power plants (Beare and Dorris, 1984), domed flight simulators (Reid, Eggemeier, and Shingledecker, 1982; Skelly and Simons, 1983), tank simulators (Whitaker, Peters, and Garinther, 1989); and the benign laboratory setting (Graham and Cook, 1984; Kilmer, Knapp, Burdsal, Borresen, Bateman, and Malzahn (1988)). In the laboratory, SWAT has been used to assess the workload associated with critical tracking and communication tasks (Reid, Shingledecker, and Eggemeier, 1981), memory tasks (Eggemeier, Crabtree, Zingg, Reid, and Shingledecker, 1982; Eggemeier and Stadler, 1984; Potter and Acton, 1985), and monitoring tasks (Notestine, 1984). Hancock and Caird (1993) reported significant increases in SWAT rating as the shrink rate of the target decreased and as the number of steps from the cursor to the target increased.

Usage in simulated flight has also been extensive (Haworth, Bivens, and Shively, 1986; Nataupsky and Abbott, 1987; Schick and Hann, 1987; Skelly and Purvis, 1985; Skelly, Reid, and Wilson, 1983; Thiessen, Lay, and Stern, 1986; Ward and Hassoun, 1990). For example, Bateman and Thompson (1986) reported that SWAT ratings increased as task difficulty increased. Their data were collected in an aircraft simulator during a tactical mission. Vickroy (1988), also using an aircraft simulator, reported that SWAT ratings increased as the amount of air turbulence increased. Fracker and Davis (1990) reported significant increases in SWAT as the number of simulated enemy aircraft increased from 1 to 3. Hancock, Williams, Manning, and Miyake (1995) reported that SWAT was highly correlated with the difficulty of a simulated flight task. However, See and Vidulich (1997) reported significant effects of target type and threat status on SWAT scores in a combat aircraft simulator. There were no significant correlations of SWAT with overall workload but two subscales correlated with peak workload (effort, $r = +0.78$; stress, $r = +0.76$).

Usage in actual flight has been extensive. For example, Pollack (1985) used SWAT to assess differences in workload between flight segments. She reported that C-130 pilots had the highest SWAT scores during the approach segment of the mission. She also reported higher SWAT ratings during the preflight segments of tactical, rather than proficiency, missions. Haskell and Reid (1987) found significant difference in SWAT ratings between flight maneuvers and also between successfully completed maneuvers and those that were not successfully completed. Gawron, et al. (1988) analyzed SWAT ratings made by the pilot and copilot four times during each familiarization and data flight: (1) during the taxi out to the runway, (2) just prior to a simulated drop, (3) just after a simulated drop, and (4) during the taxi back to the hangar. There were significant segments effects. Specifically, SWAT ratings were highest before the drop and lowest for preflight. The ratings during postdrop and postflight were both moderate.

Experience with SWAT has not been all positive, however. For example, Boyd (1983) reported that there were significant positive correlations between the three workload scales in a text-editing task. This suggests that the three dimensions of workload are not independent. This, in turn, poses a problem for use of conjoint measurement techniques. Derrick (1983) and Hart (1986) suggest that three scales may not be adequate for assessing workload. Further, Battiste and Bortolussi (1988) reported a test/retest correlation of +0.751 but also stated that, of the 144 SWAT ratings reported during a simulated B-727 flight, 59 were zero.

Corwin (1989) reported no difference between in-flight and post-flight ratings of SWAT in only two of three flight conditions. Gidcomb (1985) reported casual card sorts and urged emphasizing the importance of the card sort to SWAT raters. A computerized version of the traditional card sort was developed at the Air Force School of Aerospace Medicine. This version eliminates the tedium and dramatically reduces the time to complete the SWAT card sort. Haworth, Bivens, and Shively (1986) reported that, although the SWAT was able to discriminate control configuration conditions in a single-pilot configuration, it could not discriminate these same conditions in a pilot/copilot configuration. Wilson, Hughes, and Hassoun (1990) reported no significant differences in SWAT ratings among display formats, in contrast to pilot comments. van de Graaff (1987) reported considerable (60 points) intersubject variability in SWAT ratings during an in-flight approach task. Hill, Iavecchia, Byers, Bittner, Zaklad, and Christ (1992) reported that SWAT was not as sensitive to workload or as accepted by Army operators as NASA TLX and the Overall Workload Scale.

Vidulich and Tsang (1986) reported that SWAT failed to detect resource competition effects in dual-task performance of tracking and a directional transformation task. Vidulich (1991) reported test-retest reliability of +0.606 in SWAT ratings for tracking, choice RT, and Sternberg tasks. In addition, Rueb, Vidulich, and Hassoun (1992) reported that only one of three difficult simulated aerial refueling missions had SWAT scores above the 40 redline.

Vidulich and Pandit (1987) concluded that SWAT was not an effective measure of individual differences. This conclusion was based on no significant correlation of SWAT with any of the scales on the Jenkins Activity Survey, Rotter's Locus of Control, the Cognitive Failures Questionnaire, the Cognitive Interference Questionnaire, the Thought Occurrence Questionnaire, the California Q-sort, and the Myers-Briggs Type Indicator.

Data requirements - SWAT requires two steps to use: scale development and event scoring. Scale development requires subjects to rank, from lowest to highest workload, 27 combinations of three levels of the three workload subscales. The levels of each subscale are presented in Table 24. Reid, Eggemeier, and Nygren (1982) describe their individual differences approach to scale development. Programs to calculate the SWAT score for every combination of ratings on the three subscales are available from the Air Force Research Laboratory at Wright-Patterson Air Force Base. A user's manual is also available from the same source.

During event scoring, the subject is asked to provide a rating (1, 2, 3) for each subscale. The experimenter then maps the set of ratings to the SWAT score (1 to 100) calculated during the scale development step. Haskell and Reid (1987) suggests that the tasks to be rated be meaningful to the subjects and, further, that the ratings not interfere with performance of the task. Acton and Colle (1984) reported that the order in which the subscale ratings are presented does not affect the SWAT score. However, it is suggested that the order remain constant to minimize confusion. Eggleston and Quinn (1984) recommended developing a detailed system and operating environment description for prospective ratings.

Thresholds - Minimum value is 0, maximum value is 100. High workload is associated with the maximum value. In addition, ratings of the time, effort, and stress scales may be individually examined as workload components (Eggemeier, McGhee, and Reid, 1983).

Sources -

Acton, W.H. and Colle, H. The effect of task type and stimulus pacing rate on subjective mental workload ratings. Proceedings of the IEEE 1984 National Aerospace and Electronics Conference (pp. 818-823). Dayton, OH: IEEE; 1984.

Albery, W., Repperger, D., Reid, G., Goodyear, C., and Roe, M. Effect of noise on a dual task: subjective and objective workload correlates. Proceedings of the National Aerospace and Electronics Conference. Dayton, OH: IEEE; 1987.

Albery, W.B., Ward, S.L., and Gill, R.T. Effect of acceleration stress on human workload (Technical Report AMRL-TR-85-039). Wright-Patterson Air Force Base, OH: Aerospace Medical Research Laboratory; May 1985.

Bateman, R.P and Thompson, M.W. Correlation of predicted workload with actual workload using the subjective workload assessment technique. Proceedings of the SAE AeroTech Conference; 1986.

Battiste, V. and Bortolussi, M.R. Transport pilot workload: A comparison of two subjective techniques. Proceedings of the Human Factors Society 32nd Annual Meeting. 150-154; 1988.

Beare, A. and Dorris, R. The effects of supervisor experience and the presence of a shift technical advisor on the performance of two-man crews in a nuclear power plant simulator. Proceedings of the Human Factors Society 28th Annual Meeting. 242-246. Santa Monica, CA: Human Factors Society; 1984.

Boyd, S.P. Assessing the validity of SWAT as a workload measurement instrument. Proceedings of the Human Factors Society 27th Annual Meeting. 124-128; 1983.

Corwin, W.H. In-flight and post-flight assessment of pilot workload in commercial transport aircraft using SWAT. Proceedings of the Fifth Symposium on Aviation Psychology. 808-813; 1989.

Corwin, W.H., Sandry-Garza, D.L., Biferno, M.H., Boucek, G.P., Logan, A.L., Jonsson, J.E., and Metalis, S.A. Assessment of crew workload measurement methods, techniques, and procedures. Volume I - Process methods and results (WRDC-TR-89-7006). Wright-Patterson Air Force Base, OH; September 1989.

Crabtree, M.A. Bateman, R.P., and Acton, W.H. Benefits of using objective and subjective workload measures. Proceedings of the Human Factors Society 28th Annual Meeting (pp. 950-953). Santa Monica, CA: Human Factors Society; 1984.

Courtright J.F. and Kuperman, G. Use of SWAT in USAF system T&E. Proceedings of the Human Factors Society 28th Annual Meeting (pp. 700-703). Santa Monica, CA: Human Factors Society; 1984.

Derrick W.L. Examination of workload measures with subjective task clusters. Proceedings of the Human Factors Society 27th Annual Meeting (pp. 134-138). Santa Monica, CA: Human Factors Society; 1983.

Eggemeier, F.T. Properties of workload assessment techniques. In P.A. Hancock and N. Meshtaki (Eds.) Human mental workload (pp. 41-62). Amsterdam: North-Holland; 1988.

Eggemeier, F.T., Crabtree, M.S., and LaPoint, P. The effect of delayed report on subjective ratings of mental workload. Proceedings of the Human Factors Society 27th Annual Meeting (pp. 139-143). Santa Monica, CA: Human Factors Society; 1983.

Eggemeier, F.T., Crabtree, M.S., Zingg, J.J., Reid, G.B., and Shingledecker, C.A. Subjective workload assessment in a memory update task. Proceedings of the Human Factors Society 26th Annual Meeting, Santa Monica, CA: Human Factors Society. 643-647; 1982.

Eggemeier, F.T., McGhee, J.Z., and Reid, G.B. The effects of variations in task loading on subjective workload scales. Proceedings of the IEEE 1983 National Aerospace and Electronics Conference (pp. 1099-1106). Dayton, OH: IEEE; 1983.

Eggemeier, F.T., Melville, B., and Crabtree, M. The effect of intervening task performance on subjective workload ratings. Proceedings of the Human Factors

Society 28th Annual Meetings (pp. 954-958). Santa Monica, CA: Human Factors Society; 1984.

Eggemeier, F.T. and Stadler, M. Subjective workload assessment in a spatial memory task. Proceedings of the Human Factors Society 28th Annual Meeting (pp. 680-684). Santa Monica, CA: Human Factors Society; 1984.

Eggleston, R.G. A comparison of projected and measured workload ratings using the subjective workload assessment technique (SWAT). Proceedings of the National Aerospace and Electronics Conference, Volume 2, 827-831; 1984.

Eggleston, R.G. and Quinn, T.J. A preliminary evaluation of a projective workload assessment procedure. Proceedings of the Human Factors Society 28th Annual Meeting (pp. 695-699). Santa Monica, CA: Human Factors Society; 1984.

Fracker, M.L. and Davis, S.A. Measuring operator situation awareness and mental workload. Proceedings of the Fifth Mid-Central Ergonomics/Human Factors Conference, Dayton, OH, 23-25 May 1990.

Gawron, V.J., Schiflett, S., Miller, J., Ball, J., Slater, T., Parker, F., Lloyd, M., Travale, D., and Spicuzza, R.J. The effect of pyridostigmine bromide on inflight aircrew performance (USAFSAM-TR-87-24). Brooks Air Force Base, TX: School of Aerospace Medicine; January 1988.

Gidcomb, C. Survey of SWAT use in flight test (BDM/A-85-0630-7R.) Albuquerque, NM: BDM Corporation; 1985.

Graham, C.H. and Cook, M.R. Effects of pyridostigmine on psychomotor and visual performance (TR-84-052); September 1984.

Hancock, P.A. and Caird, J.K. Experimental evaluation of a model of mental workload. Human Factors. 35(3), 413-419; 1993.

Hancock, P.A., Williams, G., Manning, C.M., and Miyake, S. Influence of task demand characteristics on workload and performance. International Journal of Aviation Psychology. 5(1), 63-86, 1995.

Hart, S.G. Theory and measurement of human workload. In J. Seidner (Ed.) Human productivity enhancement, Vol. 1 (pp. 396-455). New York: Praeger; 1986.

Haskell, B.E., and Reid, G.B.The subjective perception of workload in low-time private pilots: A preliminary study. Aviation, Space, and Environmental Medicine. 58, 1230-1232; 1987.

Haworth, L.A., Bivens, C.C., and Shively, R.J. An investigation of single-piloted advanced cockpit and control configuration for nap-of-the-earth helicopter mission tasks. Proceedings of the 42nd Annual Forum of the American Helicopter Society. 657-671; 1986.

Hill, S.G., Iavecchia, H.P., Byers, J.C., Bittner, A.C., Zaklad, A.L., and Christ, R.E. Comparison of four subjective workload rating scales. Human Factors. 34: 429-439; 1992.

Kilmer, K.J., Knapp, R., Burdsal, C., Borresen, R., Bateman, R.P., and Malzahn, D. A comparison of the SWAT and modified Cooper-Harper scales. Proceedings of the Human Factors 32nd Annual Meeting. 155-159; 1988.

Lutmer, P.A. and Eggemeier, F.T. The effect of intervening task performance and multiple ratings on subjective ratings of mental workload. Paper presented at the 5th Mid-central Ergonomics Conference, University of Dayton, Dayton, OH; 1990.

Masline, P.J. A comparison of the sensitivity of interval scale psychometric techniques in the assessment of subjective mental workload. Unpublished masters thesis, University of Dayton, Dayton, OH; 1986.

Moroney, W.F., Biers, D.W., and Eggemeier, F.T. Some measurement and methodological considerations in the application of subjective workload measurement techniques. International Journal of Aviation Psychology. 5(1), 87-106, 1995.

Nataupsky, M. and Abbott, T.S. Comparison of workload measures on computer-generated primary flight displays. Proceedings of the Human Factors Society 31st Annual Meeting (pp. 548-552). Santa Monica, CA: Human Factors Society; 1987.

Notestine, J. Subjective workload assessment and effect of delayed ratings in a probability monitoring task. Proceedings of the Human Factors Society 28th Annual Meeting (pp. 685-690). Santa Monica, CA: Human Factors Society; 1984.

Nygren, T.E. Conjoint measurement and conjoint scaling: A users guide (AFAMRL-TR-82-22). Wright-Patterson Air Force Base, OH: Aerospace Medical Research Laboratory; April 1982.

Nygren, T.E. Investigation of an error theory for conjoint measurement methodology (763025/714404). Columbus, OH: Ohio State University Research Foundation, May 1983.

Nygren, T.E. Psychometric properties of subjective workload measurement techniques: Implications for their use in the assessment of perceived mental workload. Human Factors. 33, 17-33; 1991.

Papa, R.M. and Stoliker, J.R. Pilot workload assessment: A flight test approach. Washington, DC: American Institute of Aeronautics and Astronautics, 88-2105, 1988.

Pollack, J. Project report: an investigation of Air Force reserve pilots' workload. Dayton, OH: Systems Research Laboratory; November 1985.

Potter, S.S. and Acton, W.H. Relative contributions of SWAT dimensions to overall subjective workload ratings. Proceedings of Third Symposium on Aviation Psychology. Columbus, OH: Ohio State University; 1985.

Potter, S.S. and Bressler, J.R. Subjective workload assessment technique (SWAT): A user's guide. Wright-Patterson Air Force Base, OH: Armstrong Aerospace Medical Research Laboratory; July 1989.

Reid, G.B., Eggemeier, F., and Nygren, T. An individual differences approach to SWAT scale development. Proceedings of the Human Factors Society 26th Annual Meeting (pp. 639-642). Santa Monica, CA: Human Factors Society; 1982.

Reid, G.B., Eggemeier, F.T., and Shingledecker, C.A. In M.L. Frazier and R.B. Crombie (Eds.) Proceedings of the workshop on flight testing to identify pilot workload and pilot dynamics, AFFTC-TR-82-5 (pp. 281-288). Edwards AFB, CA: May 1982.

Reid, G.B. and Nygren, T.E. The subjective workload assessment technique: A scaling procedure for measuring mental workload. In P.A. Hancock and N. Mehtaki (Eds.) Human mental workload (pp. 185-218). Amsterdam: North Holland; 1988.

Reid, G.B., Shingledecker, C.A., and Eggemeier, F.T. Application of conjoint measurement to workload scale development. Proceedings of the Human Factors Society 25th Annual Meeting. 522-526; 1981.

Reid, G.B., Shingledecker, C.A., Nygren, T.E., and Eggemeier, F.T. Development of multidimensional subjective measures of workload. Proceedings of the IEEE International Conference on Cybernetics and Society. 403-406; 1981.

Rueb, J., Vidulich, M., and Hassoun, J.A. Establishing workload acceptability: an evaluation of a proposed KC-135 cockpit redesign. Proceedings of the Human Factors Society 36th Annual Meeting. 17-21; 1992.

Schick, F.V. and Hann, R.L. The use of subjective workload assessment technique in a complex flight task. In A.H. Roscoe (Ed.) The practical assessment of pilot workload, AGARDograph No. 282 (pp. 37-41). Neuilly-sur-Seine, France: AGARD; 1987.

See, J.E. and Vidulich, M.A. Assessment of computer modeling of operator mental workload during target acquisition. Proceedings of the Human Factors and Ergonomics Society 41st Annual Meeting. 1303-1307, 1997.

Skelly, J.J. and Purvis, B. B-52 wartime mission simulation: Scientific precision in workload assessment. Paper presented at the 1985 Air Force Conference on Technology in Training and Education. Colorado Springs, CO; April 1985.

Skelly, J.J., Reid, G.B., and Wilson, G.R. B-52 full mission simulation: Subjective and physiological workload applications. Paper presented at the Second Aerospace Behavioral Engineering Technology Conference, 1983.

Skelly, J.J. and Simons, J.C. Selecting performance and workload measures for full-mission simulation. Proceedings of the IEEE 198 National Aerospace and Electronics Conference. 1082-1085; 1983.

Thiessen, M.S., Lay, J.E., and Stern, J.A. Neuropsy-Chological workload test battery validation study (FZM 7446), Fort Worth, TX: General Dynamics; 1986.

van de Graaff, R.C. An in-flight investigation of workload assessment techniques for civil aircraft operations (NLR-TR-87119 U). Amsterdam, the Netherlands: National Aerospace Laboratory; 1987.

Vickroy, S.C. Workload prediction validation study: The verification of CRAWL predictions. Wichita, KS: Boeing Military Airplane Company; 1988.

Vidulich, M.A. The Bedford Scale: Does it measure spare capacity? Proceedings of the 6th International Symposium on Aviation Psychology. 2, 1136-1141; 1991.

Vidulich, M.A. and Pandit, P. Individual differences and subjective workload assessment: Comparing pilots to nonpilots. Proceedings of the International Symposium on Aviation Psychology. 630-636, 1987.

Vidulich, M.A. and Tsang, P.S. Techniques of subjective workload assessment: a comparison of SWAT and NASA-Bipolar methods. Ergonomics. 29(11), 1385-1398; 1986.

Vidulich, M.A. and Tsang, P.S. Absolute magnitude estimation and relative judgment approaches to subjective workload assessment. Proceedings of the Human Factors Society 31st Annual Meeting. 1057-1061; 1987.

Ward, G.F. and Hassoun, J.A. The effects of head-up display (HUD) pitch ladder articulation, pitch number location and horizon line length on unusual altitude recoveries for the F-16 (ASD-TR-90-5008). Wright-Patterson Air Force Base, OH: Crew Station Evaluation Facility; July 1990.

Warr, D.T. A comparative evaluation of two subjective workload university measures: the subjective assessment technique and the modified Cooper-Harper Rating. Masters thesis, Dayton, OH: Wright State, 1986.

Warr, D., Colle, H., and Reid, G.B. A comparative evaluation of two subjective workload measures: The subjective workload assessment technique and the modified Cooper-

Harper Scale. Paper presented at the Symposium on Psychology in the Department of Defense. USAFA, Colorado Springs, CO.; 1986.

Whitaker, L.A., Peters, L., and Garinther, G. Tank crew performance: Effects of speech intelligibility on target acquisition and subjective workload assessment. Proceedings of the Human Factors Society 33rd Annual Meeting. 1411-1413; 1989.

Wilson, G.F., Hughes, E., and Hassoun, J.A Physiological and subjective evaluation of a new aircraft display. Proceedings of the Human Factors Society 34th Annual Meeting. 1441-1443; 1990.

3.2.30 Subjective Workload Dominance

General description - The Subjective Workload Dominance (SWORD) technique uses judgment matrices to assess workload.

Strengths and limitations - SWORD is a sensitive and reliable workload measure (Vidulich, 1989). It has also been useful in projecting workload associated with various Head Up Display (HUD) formats (Vidulich, Ward, and Schueren, 1991). In addition, Tsang and Vidulich (1994) reported significant differences in SWORD ratings as a function of tracking task condition. The test-retest reliability was +0.937.

Data requirements - There are three required steps: (1) a rating scale listing all possible pairwise comparisons of the tasks performed must be completed, (2) a judgment matrix comparing each task to every other task must be filled in with each subject's evaluation of the tasks, and (3) ratings must be calculated using a geometric means approach.

Thresholds - Not stated.

Sources -

Tsang, P.S. and Vidulich, M.A. The roles of immediacy and redundancy in relative subjective workload assessment. Human Factors. 36(3), 503-513; 1994.

Vidulich, M.A. The use of judgment matrices in subjective workload assessment: the subjective workload dominance (SWORD) technique. Proceedings of the Human Factors Society 33rd Annual Meeting. 1406-1410; 1989.

Vidulich, M.A., Ward, G.F., and Schueren, J. Using the Subjective Workload Dominance (SWORD) technique for projective workload assessment. Human Factors. 33(6), 677-691; 1991.

3.2.31 Task Analysis Workload

General description - The Task Analysis/Workload (TAWL) technique requires missions to be decomposed into phases, segments, functions, and tasks. For each task a subjective matter expert rates the workload on a scale of 1 to 7. The tasks are combined into a scenario timeline and workload estimated for each point on the timeline.

Strengths and limitations - TAWL is sensitive to task workload but requires about 6 months to develop (Harris, Hill, Lysaght, and Christ, 1992). It has been used to identify workload in helicopters (Szabo and Bierbaum, 1986).

$$p=b_0+b_1N+b_2S$$

where p = utilization

b = intercept determined from regression analysis

b_1 = slope determined from regression analysis

N = number of information types

S = quantity of information in an information type

Hamilton and Cross (1993), based on seven task conditions, performance of 20 AH-64 aviators, and two analysts, reported significant correlations (+0.89 and +0.99) between the measures predicted by the TAWL model and the actual data.

Data requirements - A detailed task analysis is required. Then subject matter experts must rate the workload of each task on six channels (auditory, cognitive, kinesthetic, psychomotor, visual, and visual-aided). An IBM-PC compatible system is required to run the TAWL software. A user's guide (Hamilton, Bierbaum, and Fulford, 1991) is available.

Thresholds - Not stated.

Sources -

Hamilton, D.B., Bierbaum, C.R., and Fulford, L.A. Task Analysis/Workload (TAWL) User's guide - version 4.0 (ASI 690-330-90). Fort Rucker, AL: Anacapa Sciences; 1991.

Hamilton, D.B. and Cross, K.C. Preliminary validation of the task analysis/workload methodology (ARI RN92-18). Alexandria, VA: Army Research Institute for the Behavioral and Social Sciences; 1993.

Harris, R.M., Hill, S.G., Lysaght, R.J., and Christ, R.E. Handbook for operating the OWL & NEST technology (ARI Research Note 92-49). Alexandria, VA: United States Army Research Institute for the Behavioral and Social Sciences; 1992.

Szabo, S.M. and Bierbaum, C.R. A comprehensive task analysis of the AH-64 mission with crew workload estimates and preliminary decision rules for developing an AH-64 workload prediction model (ASI 678-204-86[B]). Fort Rucker, AL: Anacapa Sciences; 1986.

3.2.32 Utilization

General description - Utilization (p) is the probability of the operator being in a busy status (Her and Hwang, 1989).

Strengths and limitations - Utilization has been a useful measure of workload in continuous process tasks (e.g., milling, drilling, system controlling, loading, and equipment setting). It accounts for both arrival time of work in a queue and service time on that work.

Data requirements - A queuing process must be in place.

Thresholds - Minimum value is 0, maximum value is 1. High workload is associated with the maximum value.

Source -

Her, C. and Hwang, S. Application of queuing theory to quantify information workload in supervisory control systems. International Journal of Industrial Ergonomics. 4, 51-60; 1989.

3.2.33 Workload/Compensation/Interference/Technical Effectiveness

General description - The Workload/Compensation/Interference/Technical Effectiveness (WCI/TE) rating scale (see Figure 18) requires subjects to rank the sixteen matrix cells and then rate specific tasks. The ratings are converted by conjoint scaling techniques to values of 0 to 100.

Strengths and limitations - Wierwille and Connor (1983) reported sensitivity of WCI/TE ratings to three levels of task difficulty in a simulated flight task. Wierwille, Casali, Connor, and Rahimi (1985) reported sensitivity to changes in difficulty in psychomotor, perceptual, and mediational tasks. Wierwille, Rahimi, and Casali (1985) reported that WCI/TE was sensitive to variations in the difficulty of a secondary mathematical task during a simulated flight task. However, O'Donnell and Eggemeier (1986) suggest that the WCI/TE should not be used as a direct measure of workload.

Data requirements - Subjects must rank the sixteen matrix cells and then rate specific tasks. Complex mathematical processing is required to convert the ratings to WCI/TE values.

Thresholds - 0 is minimum workload, 100 is maximum workload.

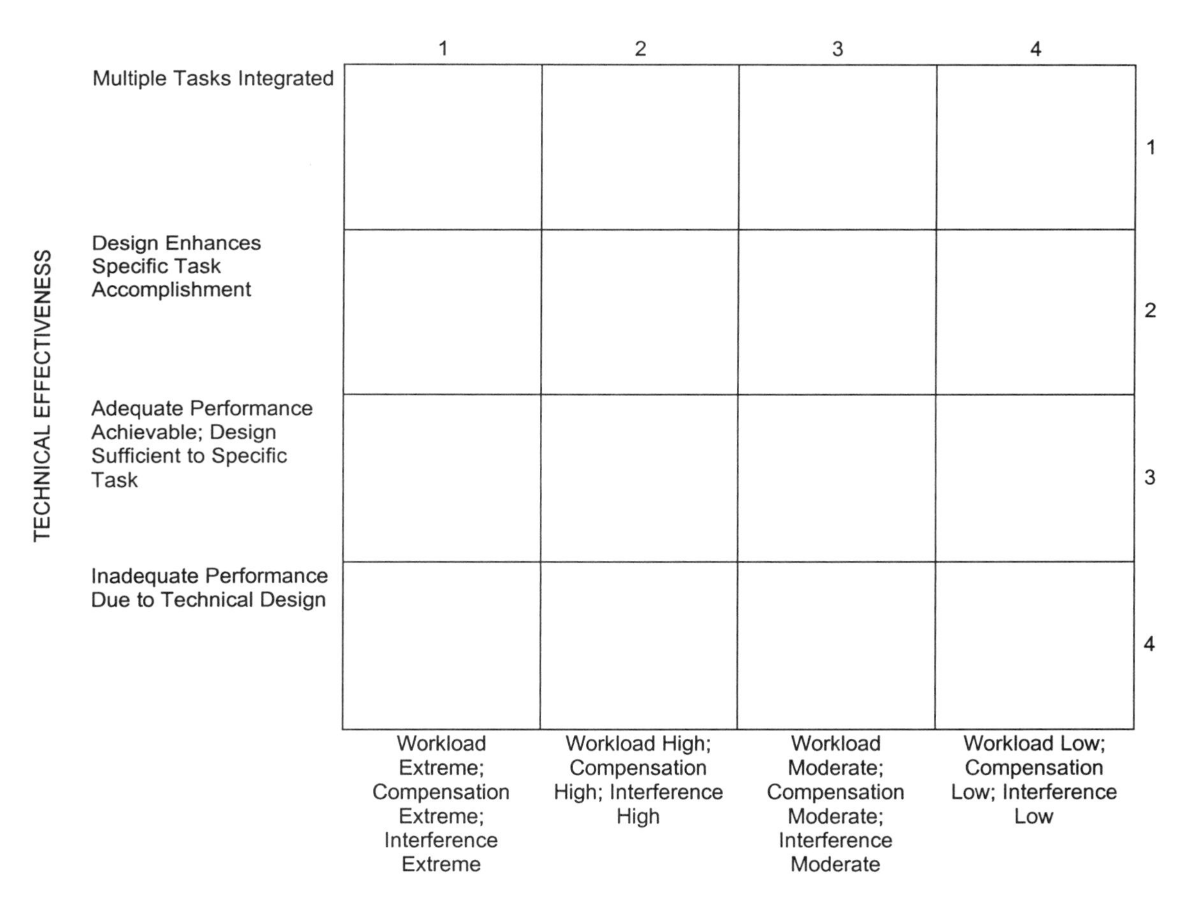

FIG. 18. WCI/TE Scale Matrix

Source -

Lysaght, R.J., Hill, S.G., Dick, A.O., Plamondon, B.D., Linton, P.M., Wierwille, W.W., Zaklad, A.L., Bittner, A.C., and Wherry, R.J. Operator workload: comprehensive review and evaluation of operator workload methodologies (Technical Report 851). Alexandria, VA. Army Research Institute for the Behavioral and Social Sciences, June 1989.

O'Donnell, R.D. and Eggemeier, F.T. Workload assessment methodology. In K.R. Boff, L. Kaufman, and J.P. Thomas (Eds.) Handbook of perception and human performance. Vol. 2, Cognitive processes and performance: New York: Wiley; 1986.

Wierwille, W.W., Casali, J.G., Connor, S.A., and Rahimi, M. Evaluation of the sensitivity and intrusion of mental workload estimation techniques. In W. Roner (Ed.) Advances in man-machine systems research. Vol. 2 (pp. 51-127). Greenwich, CT: J.A.I. Press; 1985.

Wierwille, W.W., Rahimi, M., and Casali, J.G. Evaluation of 16 measures of mental workload using a simulated flight task emphasizing mediational activity. Human Factors. 27(5), 489-502; 1985.

Wierwille, W.W. and Connor, S.A. Evaluation of twenty workload assessment measures using a psychomotor task in a motion-base aircraft simulation. Human Factors. 25, 1-16; 1983.

3.2.34 Zachary/Zaklad Cognitive Analysis

General Description - The Zachary/Zaklad Cognitive Analysis Technique requires both operational subject matter experts and cognitive scientists to identify operator strategies for performing all tasks listed in a detailed cognitive mission task analysis. A second group of subjective matter experts then rates, using 13 subscales, workload associated with performing each task.

Strengths and limitations - The method has only been applied in two evaluations, one for the P-3 aircraft (Zaklad, Deimler, Iavecchia, and Stokes, 1982), the other for F/A-18 aircraft (Zachary, Zaklad, and Davis, 1987; Zaklad, Zachary, and Davis, 1987).

Data requirements - A detailed cognitive mission timeline must be constructed. Two separate groups of subject matter experts are required, one to develop the timeline, the other to rate the associated workload.

Thresholds - Not stated.

Sources -

Zaklad, A.L., Deimler, J.D., Iavecchia, H.P., and Stokes, J. Multisensor correlation and TACCO workload in representative ASW and ASUW environments (TR-1753A). Analytics; 1982.

Zachary, W., Zaklad, A., and Davis, D. A cognitive approach to multisensor correlation in an advanced tactical environment. Proceedings of the 1987 Tri-Service Data Fusion Symposium. Fourel, MD; Johns Hopkins University; 1987.

Zaklad, A., Zachary, W., and Davis, D. A cognitive model of multisensor correlation in an advanced aircraft environment. Proceedings of the Fourth midcentral Ergonomics/Human Factors Conference, Urbana IL; 1987.

3.3 Simulation of Workload

Several digital models have been used to evaluate workload. These include: (1) Null Operation System Simulation (NOSS), (2) SAINT (Buck 1979), Modified Petri Nets (MPN) (White, MacKinnon, and Lyman, 1986), (3) task analyses (Bierbaum and Hamilton, 1990), and (4) Workload Differential Model (Ntuen and Watson, 1996).

Sources -

Bierbaum, C.R. and Hamilton, D.B. Task analysis and workload prediction model of the MH-60K mission and a comparison with UH-60A workload predictions; Volume III: Appendices H through N. (ARI Research Note 91-02). Alexandria, VA: U.S. Army Research Institute for the Behavioral and Social Sciences, October 1990.

Buck, J. Workload estimation through simulation paper presented at the Workload Program of the Indiana Chapter of the Human Factors Society. Crawfordsville, Indiana, March 31, 1979.

Ntuen, C.A. and Watson, A.R. Workload prediction as a function of system complexity. IEEE 0-8186-7493, August 1996, pp. 96-100.

White, S.A., MacKinnon, D.P., and Lyman, J. Modified Petri Net Modal sensitivity to workload manipulations (NASA-CR-177030). Moffett Field, CA: NASA Ames Research Center, 1986.

4 Measures of Situational Awareness

Situational Awareness (SA) is knowledge relevant to the task being performed. For example, pilots must know the state of their aircraft, the environment through which they are flying, and relationships between them, such as thunderstorms are associated with turbulence. It is a critical component of decision making and has been included in several models of decision making (e.g., Dorfel and Distelmaier model, 1997; see Figure 19).

SA has three levels (Endsley, 1991): Level 1, perception of the elements in the environment; Level 2, comprehension of the current situation; and Level 3, projection of future status.

There are three types of SA measures: performance (also known as query methods, Durso and Gronlund, 1999), subjective ratings, and simulation. Individual descriptions of measures of SA are provided in the following sections. A flowchart to help select the most appropriate measure is given in Figure 20.

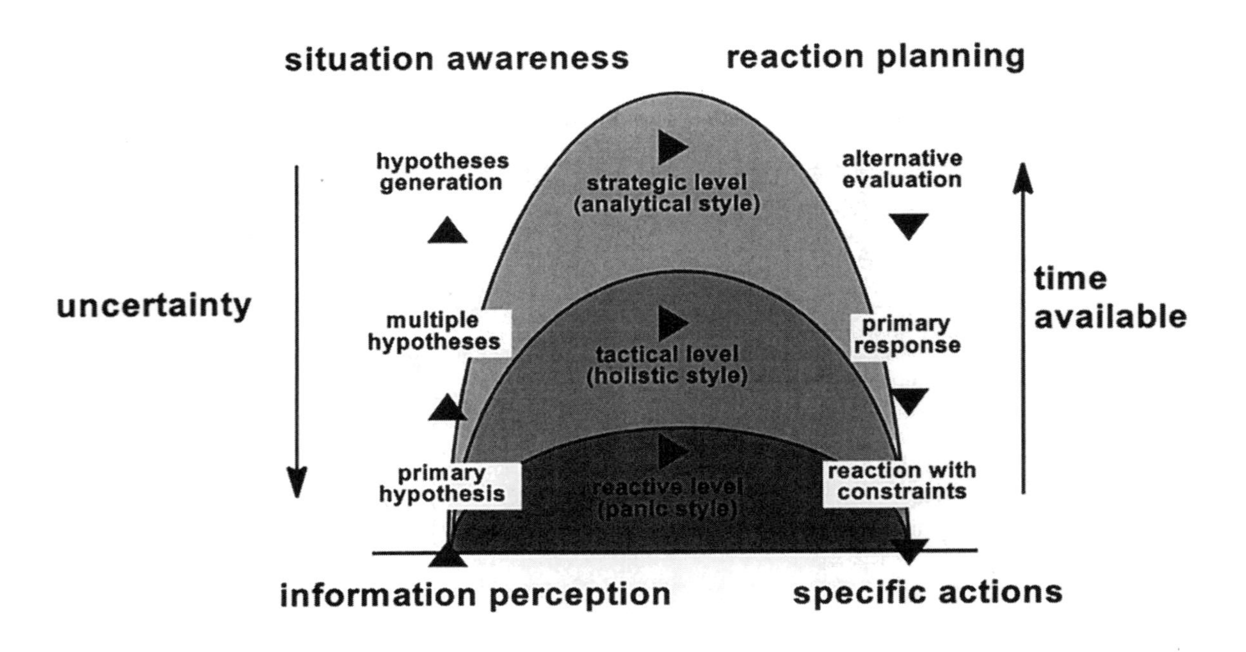

FIG. 19. Decision making under uncertainty and time pressure (Dorfel and Distelmaier, 1997, p. 2)

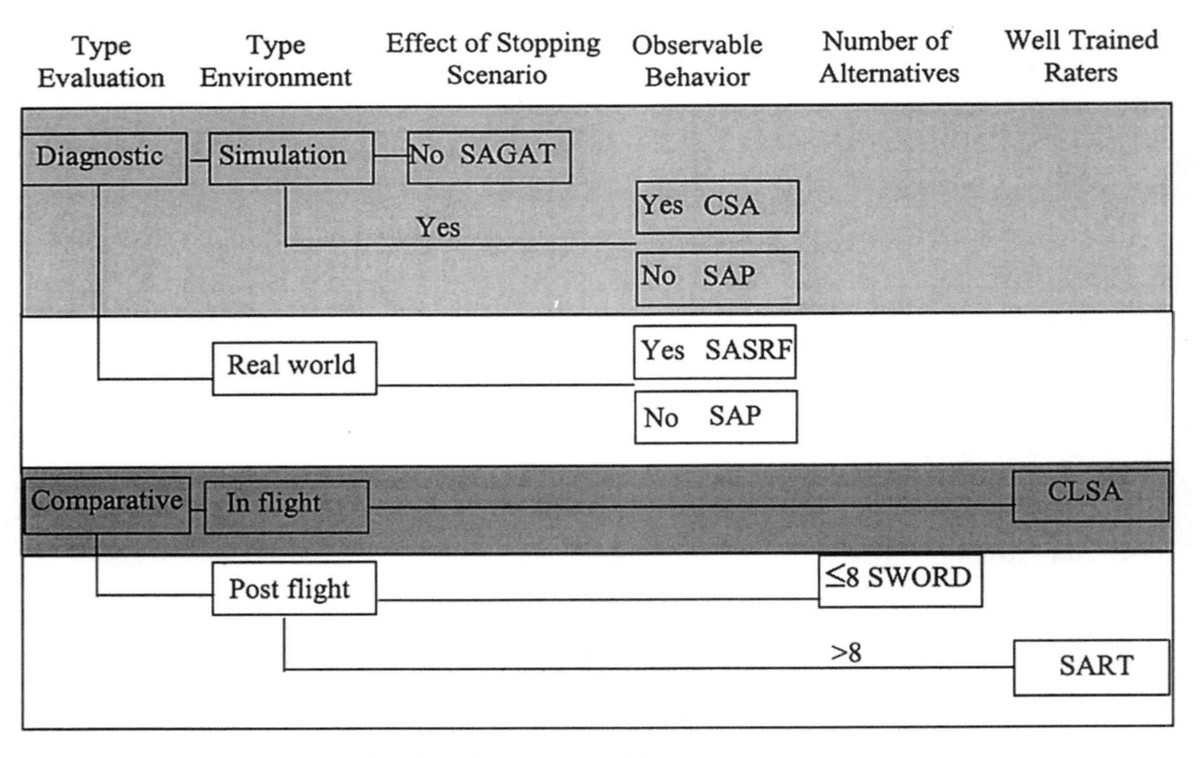

FIG. 20. Guide to selecting a SA measure

Sources -

Dorfel, G. and Distelmaier, H. Enhancing Situational Awareness by knowledge-based user interfaces. Proceedings of the 2nd Annual Symposium and Exhibition on Situational Awareness in the Tactical Air Environment. Patuxent River, MD: Naval Air Warfare Center, 1997.

Durso, F.T. and Gronlund, S.D. Situation Awareness. In F.T. Durso, R.S. Nickerson, R.W. Schvaneveldt, S.T. Dumais, D.S. Lindsay, and M.T.H. Chi (Eds). Handbook of applied cognition. New York: John Wiley and Sons, 1999.

Endsley, M.R. Situation Awareness in dynamic systems. In R.M. Taylor (Ed.) Situational awareness in dynamic systems (IAM Report 708). Farnborough, UK: Royal Air Force Institute of Aviation Medicine, 1991.

4.1 Performance Measures of SA

4.1.1 Situational Awareness Global Assessment Technique

General description - The most well known measure of SA is the Situational Awareness Global Assessment Technique (SAGAT) (Endsley, 1988a). SAGAT was designed around real-time, human-in-the-loop simulation of a military cockpit but could be generalized to other systems. Using SAGAT, the simulation is stopped at random times and the operators are asked questions to determine their SA at that particular point in time. Subjects' answers are compared with the correct answers that have been simultaneously collected in the computer database. "The comparison of the real and perceived situation provides an objective measure of SA" (Endsley, 1988b, p. 101).

This same technique could be used with any complex system that is simulated, be it a nuclear power plant control room or the engine room of a ship. In addition, if an operational system is properly instrumented, SAGAT is also applicable in this environment. SAGAT uses a graphical computer program for the rapid presentation of queries and data collection. In addition to possessing a high degree of face validity, the SAGAT technique has been tested in several studies, which demonstrated: (1) empirical validity (Endsley, 1989, 1990b) - the technique of freezing the simulation did not impact subject performance and subjects were able to reliably report SA knowledge for up to six minutes after a freeze without memory decay problems; (2) predictive validity (Endsley, 1990b) - linking SAGAT scores to subject performance; and (3) content validity (Endsley, 1990a) - showing appropriateness of the queries used (for an air-to-air fighter cockpit).

Other researchers have simply asked subjects to indicate locations of objects on maps. This technique is sometimes refer to as mini-SAGAT or snapshots. It has been used in a wide variety of military and civilian tasks. For example, firefighters have been asked to mark the location of a fire on a map. SA is measured as the deviation between the actual fire and the indicated location on the map (Artman, 1999).

Strengths and limitations - SAGAT provides unbiased objective measures of SA across all of the operators' SA requirements that can be computed in terms of errors or percent correct and can be treated accordingly. However, Sarter and Woods (1991) suggest that SAGAT does not measure SA but rather measures what pilots can recall. Further, Fracker and Vidulich (1991) identified two major problems with the use of explicit measures of SA, such as SAGAT: (1) decay of information and (2) inaccurate beliefs.

Fracker (1989) stated from a simulated air threat study "the present data encourage the use of memory probes to measure situation awareness." Fracker (1991) evaluated SA measures in a simulated combat task. Test-retest correlations were significant for identity accuracy, identity latency, envelope sensitivity, and kill probability but not for location error and avoidance failure. Only identity latency was significantly correlated with kill probability. This correlation was used to assess criterion validity.

Construct validity was evaluated from correlations of SA metrics with avoidance failure. Three correlations were significant: 1) identity latency, 2) envelope sensitivity, and 3) kill probability. Two correlations were not significant: 1) location error and 2) identity accuracy. There were three significant correlations among SA metrics: 1) identity accuracy and location error, 2) identity accuracy and identity latency, and 3) envelope sensitivity and latency. A 0.10 alpha was used to determine significance. Three correlations were not significant: 1) identity latency and location error, 2) envelope sensitivity and location error, and 3) envelope sensitivity and identify accuracy. Endsley (1995) did not find any significant difference in SAGAT scores as a function of time (0 to 400s) between event and response nor did she see any performance decrements in a piloting task between stopping the simulation and not stopping the simulation.

Data requirements - The proper queries must be identified prior to the start of the experiment.

Thresholds - Tolerance limits for acceptable deviance of perceptions from real values on each parameter should be identified prior to the start of the experiment.

Sources -

Artman, H. Situation awareness and co-operation within and between hierarchical units in dynamic decision making. Ergonomics, 1999, 42(11), 1404-1417.

Endsley, M.R. Situational awareness global assessment technique (SAGAT). Proceedings of the National Aerospace and Electronics Conference. 789-795; 1988a.

Endsley, M.R. Design and evaluation for situation awareness enhancement. Proceedings of the 32nd Annual Meeting of the Human Factors Society. 97-101; 1988b.

Endsley, M.R. A methodology for the objective measurement of pilot situation awareness. Presented at the AGARD Symposium on Situation Awareness in Aerospace Operations. Copenhagen, Denmark; October 1989.

Endsley, M.R. Situation awareness in dynamic human decision making: theory and measurement (NORDOC 90-49). Hawthorne, CA: Northrop Corporation, 1990a.

Endsley, M.R. Predictive utility of an objective measure of situation awareness. Proceedings of the Human Factors Society 34th Annual Meeting (pp. 41-45), Santa Monica, CA: Human Factors Society, 1990b.

Endsley, M.R. Toward a theory of situational awareness in dynamic systems. Human Factors. 1995, 37(1); 32-64.

Fracker, M.L. Attention allocation in situation awareness. Proceedings of the Human Factors Society 33rd Annual Meeting, 1989.

Fracker, M.L. Measures of Situation Awareness: An experimental evaluation (AL-TR-1991-0127). Wright-Patterson Air Force Base, OH: October 1991.

Fracker, M.L. and Vidulich, M.A. Measurement of situation awareness: a brief review. In R.M. Taylor (Ed.) Situational awareness in dynamic systems (IAM Report 708). Farnborough, UK: Royal Air Force Institute of Aviation Medicine, 1991.

Sarter, N.B. and Woods, D.D. Situational awareness: a critical but ill-defined phenomenon. International Journal of Aviation Psychology. 1(1), 45-57; 1991.

4.1.2 Situational Awareness Linked Instances Adapted to Novel Tasks

General description - The Situational Awareness Linked Instances Adapted to Novel Tasks (SALIANT) was developed to measure team SA. The SALIANT methodology requires five phases: 1) identify team SA behaviors (see Table 26), 2) develop scenarios, 3) define acceptable responses, 4) write a script, and 5) create a structured form with columns for scenario and responses.

Strengths and limitations - SALIANT has been validated using twenty undergraduate students in a four-hour tabletop helicopter simulation. Interrater reliability was r = +0.94. There were significant correlations between SALIENT score and communication frequency (r = +0.74), between SALIENT score and performance (r = +0.63). There were no significant correlations between SALIANT score and the teams' shared mental model (r = -0.04). Additional validation data aere available in Muniz, Salas, Stout, and Bowers (1998).

TABLE 26.
Generic Behavioral Indicators of Team SA (Muniz, Stout, Bowers, and Salas, 1998)

Demonstrated Awareness of Surrounding Environment
Monitored environment for changes, trends, abnormal conditions
Demonstrated awareness of where he/she was
Recognized Problems
Reported problems
Located potential sources of problem
Demonstrated knowledge of problem consequences
Resolved discrepancies
Noted deviations
Anticipated a need for action
Recognized a need for action
Anticipated consequences of actions and decisions
Informed others of actions taken
Monitored actions
Demonstrated knowledge of tasks
Demonstrated knowledge of tasks
Exhibited skill time sharing attention among tasks
Monitored workload
Shared workload within station
Answered questions promptly
Demonstrated awareness of information
Communicated important information
Confirmed information when possible
Challenged information when doubtful
Re-checked old information
Provided information in advance
Obtained information of what is happening
Demonstrated understanding of complex relationship
Briefed status frequently

Data requirements - Although the generic behaviors in Table 26 can be used, scenarios, responses, scripts, and report forms must be developed for each team task.

Thresholds - Not stated.

Sources –

Muniz, E.J., Salas, E., Stout, R.J., and Bowers, C.A. The validation of a team Situational Awareness Measure. Proceedings for the Third Annual Symposium and Exhibition on Situational Awareness in the Tactical Air Environment. Patuxent River, MD: Naval Air Warfare Center Aircraft Division, 1998, 183-190.

Muniz, E.J., Stout, R.J., Bowers, C.A., and Salas, E. A methodology for measuring team situational awareness: Situational Awareness Linked Indicators Adapted to Novel Tasks (SALIANT). The First Annual Symposium/Business Meeting of the Human Factors and Medicine Panel on Collection Crew Performance in Complex Systems, Edenburg, United Kingdom, 1998.

4.1.3 Temporal Awareness

General description - Temporal awareness has been defined as "the ability of the operator to build a representation of the situation including the recent past ands the near future" (Grosjean and Terrier, 1999, p. 1443). It has been hypothesized to be critical to process management tasks.

Strengths and limitations - Temporal awareness has been measured as the number of temporal and ordering errors in a production line task, number of periods in which temporal constraints were adhered to, and the temporal landmarks reported by the operator to perform his or her task. Temporal landmarks include relative ordering of production lines and clock and mental representation of the position of production lines.

Data requirements - Correct time and order must be defined with tolerances for error data. Temporal landmarks must be identified during task debriefs.

Thresholds - Not stated.

Source -

Grosjean, V. and Terrier, P. Temporal awareness: pivotal in performance? Ergonomics, 1999, 42(11), 1443-1456.

4.2 Subjective Measures of SA

Subjective measures of SA share many of the advantages and limitations of subjective measures of workload discussed in Section 3.2. Advantages include: inexpensive, easy to administer, and high face validity. Disadvantages include: inability to measure what the subject cannot describe well in words and requirement for well-defined questions.

4.2.1 China Lake Situational Awareness

General description - The China Lake Situational Awareness (CLSA) is a five-point rating scale (see Table 27) based on the Bedford Workload Scale. It was designed at the Naval Air Warfare Center at China Lake to measure SA in flight (Adams, 1998).

Strengths and limitations –

Data requirements – Points in the flight during which the aircrew are asked to rate their SA using CLSA Rating Scale must not compromise safety.

TABLE 27.
China Lake SA Rating Scale

SA SCALE VALUE	CONTENT
VERY GOOD 1	• Full knowledge of a/c energy state/tactical environment/mission; • Full ability to anticipate/accommodate trends
GOOD 2	• Full knowledge of a/c energy state/tactical environment/mission; • Partial ability to anticipate/accommodate trends; • No task shedding
ADEQUATE 3	• Full knowledge of a/c energy state/tactical environment/mission; • Saturated ability to anticipate/accommodate trends; • Some shedding of minor tasks
POOR 4	• Fair knowledge of a/c energy state/tactical environment/mission; • Saturated ability to anticipate/accommodate trends; • Shedding of all minor tasks as well as many not essential to flight safety/mission effectiveness
VERY POOR 5	• Minimal knowlege of a/c energy state/tactical environment/mission; • Oversaturated ability to anticipate/accommodate trends; • Shedding of all tasks not absolutely essential to flight safety/mission effectiveness

Thresholds – 1 (very good) to 5 (very poor).

Source –

Adams, S. Practical considerations for measuring Situational Awareness. Proceedings for the Third Annual Symposium and Exhibition on Situational Awareness in the Tactical Air Environment, 1998, 157-164.

4.2.2 Crew Situational Awareness

General description - Mosier and Chidester (1991) developed a method for measuring situational awareness of air transport crews. Expert observers rate crew coordination performance and identify and rate performance errors (type 1, minor errors; type 2, moderately severe errors; and type 3, major, operationally significant errors). The experts then develop information transfer matrices identifying time and source of item requests (prompts) and verbalized responses. Information is then classified into decision or nondecision information.

Strengths and limitations - The method was sensitive to type of errors and decision prompts.

Data requirements - The method requires open and frequent communication among aircrew members. It also requires a team of expert observers to develop the information transfer matrices.

Thresholds - Not stated.

Source-

Mosier, K.L. and Chidester, T.R. Situation assessment and situation awareness in a team setting. In R.M. Taylor (Ed.) Situation awareness in dynamic systems (IAM Report 708). Farnborough, UK: Royal Air Force Institute of Aviation Medicine, 1991.

4.2.3 Human Interface Rating and Evaluation System

General description - The Human Interface Rating and Evaluation System (HiRes) is a generic judgment-scaling technique developed by Budescu, Zwick, and Rapoport (1986).

Strengths and limitations - HiRes has been used to evaluate SA (Fracker and Davis, 1990). These authors reported a significant effect of the number of enemy aircraft in a simulation and HiRes rating.

Data requirements - HiRes ratings are scaled to sum to 1.0 across all the conditions to be rated.

Thresholds - 0 to 1.0

Sources -

Budescu, D.V., Zwick, R., and Rapoport, A. A comparison of the eigenvalue and the geometric mean procedure for ratio scaling. Applied Psychological Measurement. 10, 69-78; 1986.

Fracker, M.L. and Davis, S.A. Measuring operator Situation Awareness and mental workload. Proceedings of the fifth Mid-Central Ergonomics/Human Factors Conference, Dayton, OH, 23-25 May 1990.

4.2.4 Situational Awareness Rating Technique

General description - An example of a subjective measure of SA is the Situational Awareness Rating Technique (SART) (Taylor, 1990). SART is a questionnaire method that concentrates on measuring the operator's knowledge in three areas: (1) demands on attentional resources, (2) supply of attentional resources, and (3) understanding of the situation (see Figure 21 and Table 28). The reason that SART measures three different components (there is also a 10-dimensional version) is that the SART developers feel that, like workload, SA is a complex construct; therefore, to measure SA in all its aspects, separate measurement dimensions are required. Because information processing and decision making are inextricably bound with SA (since SA involves primarily cognitive rather than physical workload), SART has been tested in the context of Rasmussen's Model of skill-, rule-, and knowledge-based behavior. Selcon and Taylor (1989) conducted separated studies looking at the relationship between SART and rule- and knowledge-based decisions, respectively. The results showed that SART ratings appear

		LOW						HIGH
		1	2	3	4	5	6	7
Demand	Instability of Situation							
	Variability of Situation							
	Complexity of Situation							
Supply	Arousal							
	Spare Mental Capacity							
	Concentration							
	Division of Attention							
Under	Information Quantity							
	Information Quality							
	Familiarity							

FIG. 21. SART Scale

TABLE 28.
Definitions of SART Rating Scales

Demand on Attentional Resources
Instability: Likelihood of situation changing suddenly.
Complexity: Degree of complication of situation.
Variability: Number of variables changing in situation.

Supply of Attentional Reources
Arousal: Degree of readiness for activity.
Concentration: Degree of readiness for activity.
Division: Amount of attention in situation.
Space Capacity: Amount of attentional left to spare for new variables.

Understanding of the Situation
Information Quantity: Amount of information received and understood.
Information Quality: Degree of goodness of information gained.

From Taylor and Selcon (1991, p. 10)

to provide diagnosticity in that they were significantly related to performance measures of the two types of decision making. Early indications are that SART is tapping the essential qualities of SA, but further validation studies are required before this technique is commonly used.

Strengths and limitations - SART is a subjective measure and, as such, suffers from the inherent reliability problems of all subjective measures. The strengths are that SART is easily administered and was developed in three logical phases: (1) scenario generation, (2) construct elicitation, and (3) construct structure validation (Taylor, 1989). SART has been prescribed for comparative system design evaluation (Taylor and Selcon, 1991). SART is sensitive to differences in performance of aircraft attitude recovery tasks and learning comprehension tasks (Selcon and Taylor, 1991; Taylor and Selcon, 1990). SART is also sensitive to pilot experience (Selcon, Taylor, and Koritsas, 1991).

Vidulich, McCoy, and Crabtree (1995) reported increased difficulty of a PC-based flight simulation increased the Demand Scale on the SART but not the Supply or Understanding Scales. Providing additional information increased the Understanding Scale but not the Demand or Supply Scales.

However, Taylor and Selcon (1991) state "There remains considerable scope for scales development, through description improvement, interval justification and the use of conjoint scaling techniques to condense multi-dimensional ratings into a single SA score" (p. 11). These authors further state that "The diagnostic utility of the Attentional Supply constructs has yet to be convincingly demonstrated" (p.12).

Selcon, Taylor, and Shadrake (1992) used SART to evaluate the effectiveness of visual, auditory, or combined cockpit warnings. Demand was significantly greater for the visual than for the auditory or the combined cockpit warnings. Neither supply nor understanding ratings were significantly different across these conditions. Similarly, Selcon, Hardiman, Croft, and Endsley (1996) reported significantly higher SART scores when a launch success zone display was available to pilots during a combat aircraft simulation. Understanding, information quantity, and information quality were also significantly higher with this display. There were no effects on Demand or Supply ratings.

Taylor, Selcon, and Swinden (1995) reported significant differences in both 3-D and 10-D SART ratings in an Air Traffic Control (ATC) task. Only three scales of the 10-D SART did not show significant effects as the number of aircraft being controlled changed (Information Quantity, Information Quality, and Familiarity).

See and Vidulich (1997) reported significant effects of target and display type on SART. The combined SART as well as the supply and understanding scales were significantly correlated to workload (r = -0.73, -0.75, and -0.82, respectively).

Data requirements - Data are on an ordinal scale; interval or ratio properties cannot be implied.

Thresholds - The data are on an ordinal scale and must be treated accordingly when statistical analysis is applied to the data. Non-parametric statistics may be the most appropriate analysis method.

Sources -

See, J.E. and Vidulich, M.A. Assessment of computer modeling of operator mental workload during target acquisition. Proceedings of the Human Factors and Ergonomics Society 41st Annual Meeting. 1303-1307, 1997.

Selcon, S.J., Hardiman, T.D., Croft, D.G., and Endsley, M.R. A test-battery approach to cognitive engineering: to meta-measure or not to meta-measure, that is the question! Proceedings of the Human Factors and Ergonomics Society 40th Annual Meeting. 228-232, 1996.

Selcon, S.J. and Taylor, R.M. Evaluation of the situational awareness rating technique (SART) as a tool for aircrew systems design. AGARD Conference Proceedings No. 478. Neuilly-sur-Seine, France; 1989.

Selcon, S.J. and Taylor, R.M. Decision support and situational awareness. In R.M. Taylor (Ed.) situational awareness in dynamic systems (IAM Report 708). Farnborough, UK: Royal Air Force Institute of Aviation Medicine, 1991.

Selcon, S.J., Taylor, R.M. and Koritsas, E. Workload or situational awareness?: TLX vs. SART for aerospace systems design evaluation. Proceedings of the Human Factors Society 35th Annual Meeting. 62-66, 1991.

Selcon, S.J., Taylor, R.M., and Shadrake, R.A. Multi-modal: pictures, words, or both? Proceedings of the Human Factors Society 36th Annual Meeting. 57-61, 1992.

Taylor, R.M. Situational awareness rating technique (SART): the development of a tool for aircrew systems design. Proceedings of the NATO Advisory Group for Aerospace Research and Development (AGARD) Situational Awareness in Aerospace Operations Symposium (AGARD-CP-478); October 1989.

Taylor, R.M. Situational awareness: aircrew constructs for subject estimation (IAM-R-670); 1990.

Taylor, R.M. and Selcon, S.J. Understanding situational awareness. Proceedings of the Ergonomics Society's 1990 Annual Conference. Leeds, England; 1990.

Taylor, R.M. and Selcon, S.J. Subjective measurement of situational awareness. In R.M. Taylor (Ed.) situational awareness in dynamic systems (IAM Report 708). Farnborough, UK: Royal Air Force Institute of Aviation Medicine, 1991.

Taylor, R.M., Selcon, S.J., and Swinden, A.D. Measurement of situational awareness and performance: a unitary SART index predicts performance on a simulated ATC task. Proceedings of the 21st Conference of the European Association for Aviation Psychology. Chapter 41, 1995.

Vidulich, M.A., McCoy, A.L., and Crabtree, M.S. The effect of a situation display on memory probe and subjective situational awareness metrics. Proceedings of the 8th International Symposium on Aviation Psychology, 2, 765-768, 1995.

4.2.5 Situational Awareness Subjective Workload Dominance

General descriptions - The Situation Awareness Subjective Workload Dominance Technique (SA SWORD) uses judgment matrices to assess SA.

Strengths and limitations - Fracker and Davis (1991) evaluated alternate measures of SA on three tasks: 1) flash detection, 2) color identification, and 3) location. Ratings were made of awareness of object location, color, flash, and mental workload. All ratings were collected using a paired comparisons technique. Color inconsistency decreased SA and increased workload. Flash probability had no significant effects on the ratings.

Data requirements - There are three required steps: 1) a rating scale listing all possible pairwise comparisons of the tasks performed must be completed, 2) a judgment matrix comparing each task to every other task must be filled in with each subject's evaluation of the tasks, and 3) ratings must be calculated using a geometric means approach.

Thresholds - Not stated.

Source -

Fracker, M.L. and Davis, S.A. Explicit, implicit, and subjective rating measures of Situation Awareness in a monitoring task (AL-TR-1991-0091). Wright-Patterson Air Force Base, OH; October 1991.

4.2.6 Situational Awareness Supervisory Rating Form

General descriptions - Carretta, Perry, and Ree (1966) developed the Situational Awareness Supervisory Rating Form to measure the SA capabilities of F-15 pilots. The form has 31 items that range from general traits to tactical employment (Table 29).

Strengths and limitations - Carretta et al. (1996) reported that 92.5% of the variance in peer and supervisory ratings were due to one principal component. The best predictor of the form rating was flying experience ($r = +0.704$).

Data requirements - Supervisors and peers must make the rating.

Source -

Carretta, T.R., Perry, D.C., and Ree, M.J. Prediction of situational awareness in F-15 pilots. International Journal of Aviation Psychology. 6(1), 21-41; 1996.

4.3 Simulation

Shively, Brickner, and Silberger (1997) developed a computational model of SA. The model has three components: 1) situational elements, i.e., parts of the environment that define the situation, 2) context-sensitive nodes, i.e., semantically-related collections of situational elements, and 3) a regulatory mechanism that assesses the situational elements for all nodes.

Source –

Shively, R.J., Brickner, M., and Silbiger, J. A computational model of Situational Awareness instantiated in MIDAS, http://caffeine.arc.nasa.gov/midas/Tech_Reports.html, 1997.

TABLE 29.
Situational Awareness Supervisory Rating Form

Rater ID#: ______ Pilot ID#: ______						
Item Ratings	**Relative Ability Compared with Other F-15C Pilots**					
	Acceptable		**Good**		**Outstanding**	
	1	**2**	**3**	**4**	**5**	**6**
General traits 1. Discipline 2. Decisiveness 3. Tactical knowledge 4. Time-sharing ability 5. Reasoning ability 6. Spatial ability 7. Flight management						
Tactical game plan 8. Developing plan 9. Executing plan 10. Adjusting plan on the fly						
System operation 11. Radar 12. TEWS 13. Overall weapons system proficiency						
Communication 14. Quality (brevity, accuracy, timeliness, completeness) 15. Ability to effectively use comm information						
Information interpretation 16. Interpreting VSD 17. Interpreting RWR 18. Ability to effectively use AWACS/GCI 19. Integrating overall information (cockpit displays, wingman comm, controller comm) 20. Radar sorting 21. Analyzing engagement geometry 22. Threat prioritization						
Tactical employment-BVR weapons 23. Targeting decision 24. Fire-point selection						
Tactical employment-visual maneuvering 25. Maintain track of bogeys/friendlies 26. Threat evaluation 27. Weapons employment						
Tactical employment-general 28. Assessing offensiveness/defensiveness 29. Lookout (VSD interpretation, RWR monitoring, visual lookout) 30. Defensive reaction (chaff, flares, maneuvering, etc.) 31. Mutual support						
Overall situational awareness[a]						
Overall fighter ability						

[a] Items 1 through 31 are used for supervisory ratings. The overall fighter ability and situational awareness items are completed by both supervisors and peers. (Carretta, Perry, and Ree, 1996, pp. 40-41).

Glossary of Terms

AGARD	Advisory Group for Research and Development
AGL	Above Ground Level
FOM	Figure of Merit
FOV	Field of View
HSI	Horizontal Situation Indicator
IMC	Instrument Meteorological Conditions
ISI	Inter-stimulus interval
rmse	Root mean squared error
RT	reaction time
SD	standard deviation
STRES	Standardized Tests for Research with Environmental Stressors
VCE	Vector Combination of Errors
VMC	Visual Meterological Conditions

Author Index

C

N

Q

R

S

T

U

V

W

Y

Z

Subject Index

Y

Z